LINEAR ALGEBRA LABS with MATLAB®

Third Edition

DAVID R. HILL
DAVID E. ZITARELLI

PEARSON
Prentice Hall

Upper Saddle River, NJ 07458

Acquisitions Editor: George Lobell
Supplement Editor: Jennifer Brady
Assistant Managing Editor: John Matthews
Production Editor: Jeffrey Rydell
Supplement Cover Manager: Paul Gourhan
Supplement Cover Designer: Joanne Alexandris
Manufacturing Buyer: Ilene Kahn

Pearson Prentice Hall
Pearson Education, Inc.
Upper Saddle River, NJ 07458

Printed in the United States of America

10 9 8 7 6

ISBN 0-13-143274-5

Pearson Education Ltd., *London*
Pearson Education Australia Pty. Ltd., *Sydney*
Pearson Education Singapore, Pte. Ltd.
Pearson Education North Asia Ltd., *Hong Kong*
Pearson Education Canada, Inc., *Toronto*
Pearson Educación de Mexico, S.A. de C.V.
Pearson Education—Japan, *Tokyo*
Pearson Education Malaysia, Pte. Ltd.

Preface

This work has two primary parts. One consists of the 13 LABS and three Projects in this manual. The other is a set of instructional M-files that harness the power of the software package MATLAB ® to render it appropriate for an educational setting.[1] (See page x for an overview of the instructional routines.)

The LABS and Projects are meant to supplement a standard sophomore level course in linear algebra. They follow the general outline for such a course, introducing instructional routines and appropriate MATLAB commands to solve problems related to each concept.

Our primary goal is to use the laboratory experiences to aid in understanding the basic ideas of linear algebra. As such we use instructional M-files that provide a tool kit [2] for working with linear algebra without the need for programming in the MATLAB command set. Although no programming background is assumed, those students with computing skills can further enhance their skills within MATLAB . We have found that students initially rely on the tool kit, but many quickly begin to use MATLAB commands directly, even though we provide little formal instruction in this area. We recommend an instructional approach that integrates the language and terminology of computing within the lecture format. In addition, when possible and appropriate, computer demonstrations and experiments should be used in lectures.

Three of the LABS are different from the others. LAB 5 examines sets with addition and scalar multiplication and investigates the defining properties of a vector space in a pedagogical way. LAB 8 presents the defining properties of the determinant in such a way that a considerable amount of class time can be saved on this topic. Also, LAB 11 presents an independent supplement to the standard classroom coverage of linear transformations by examining the geometry of plane linear transformations. New Section 11.2 introduces homogeneous coordinates to incorporate translations.

The LABS are not self contained. Except for LABS 8 and 11, they assume that the material has already been presented in the classroom. Sometimes, however, it is expedient to discuss a topic using a fresh, computational approach.

New material has been added to this third addition, both in the LABS and in the accompanying instructional M-files. The modifications to the LABS provide a number of alternate

[1]The MATLAB software must be purchased separately from this work.

[2]The tool kit of instructional M-files is available free of charge. The files are available at http://www.prenhall.com through the Companion Web Sites tab. This tool kit was developed by the authors, not The MathWorks, and is to be used strictly for instructional purposes. The tool kit is compatible with MATLAB 6.5 or higher for Windows, and the Student Edition of MATLAB for Microsoft Windows. Most features of the instructional routines are compatible with corresponding MacIntosh and workstation versions.

approaches to topics some of which use more graphically oriented M-files to provide visualization of concepts. Many of the instructional M-files have been enhanced to take advantage of the graphical user interface (GUI) available in MATLAB . In addition we have included instructional files that use the Symbolic Math Toolbox. These sections can be omitted without loss of continuity if this toolbox is not available. A detailed list of new features is on page viii and a short description of all the instructional files is on page x. A full description of the instructional files is available by printing **alldesc.txt** that accompanies the tool kit of instructional files.

We extend our sincere gratitude to the National Science Foundation (ILI #DMS-9051282) for providing the funds for implementing a mathematics laboratory at Temple University. This facility provided the educational arena necessary to develop the laboratory materials and extend our instructional M-files for MATLAB from 1990 to 1993. We thank the many students who were patient with and receptive to using the laboratory to aid in the development and understanding of the concepts of linear algebra.

A special thanks to our colleague Dr. Nicholas Macri for his valuable assistance in designing and preparing this manual.

David R. Hill
David E. Zitarelli

May, 2003

Please address comments or inquiries to the authors using the following regular mail or electronic mail addresses or call us direct.

Dr. David R. Hill
Mathematics Department
Temple University
Philadelphia, Pa. 19122 USA

215 – 204 – 1654
hill@math.temple.edu

Dr. David E. Zitarelli
Mathematics Department
Temple University
Philadelphia, Pa. 19122 USA

215 – 204 – 7844
david.zitarelli@temple.edu

Introduction to MATLAB and Some of its Features[3]

MATLAB is a versatile piece of software with linear algebra capabilities as its core. MATLAB stands for **MAT**rix **LAB**oratory. It incorporates portions of professionally developed projects of quality computer routines for linear algebra computation. The MATLAB kernel is written in the C language but many of the routines are implemented in the MATLAB language.

Once you initiate the MATLAB software you will see the MATLAB logo appear and the MATLAB prompt ≫. The prompt ≫ indicates that MATLAB is awaiting a command. In LAB 1 we describe how to enter matrices into MATLAB and explain some commands. However, there are certain MATLAB features you should be aware of before you begin to use MATLAB .

MATLAB has a wide range of capabilities. In this course we will use only a small portion of its features. We will find that MATLAB 's command structure is very close to the way we write algebraic expressions and linear algebra operations. The names of many MATLAB commands closely parallel those of the operations and concepts of linear algebra. The MATLAB software provides immediate on screen descriptions using the help command. Typing

help

displays a list of MATLAB subdirectories and alternate directories containing files corresponding to commands and data sets. Typing **help name**, where **name** is the name of a command, gives information on the specific command. In some cases the description displayed goes much further than we need for this course. Hence you may not fully understand all of the description displayed by help. For example, typing

help colon

(Note: there is a space between 'help' and 'colon'.) reveals several uses that we will describe in LAB 1. However, typing

help :

reveals a list of commands of little interest in a first course in linear algebra.

[3]The descriptions apply to MATLAB for Windows. For other versions check your MATLAB users guide.

- Entering Commands.

  It is recommended that all commands in MATLAB be typed in lower case letters.

- Starting Execution of a Command.

  After you have typed a command name and any arguments or data required you must press ENTER before it will begin to execute.

- The Command Stack.

  As you enter commands MATLAB saves a number of the most recent commands in a stack. Previous commands saved on the stack can be recalled using the **up arrow** key. The number of commands saved on the stack varies depending upon the length of the commands and other factors.

- Editing Commands.

  If you make an error or mistype something in a command, you can use the **left arrow** and **right arrow** keys to position the cursor for corrections. The **home** key moves the cursor to the beginning of a command and the **end** key moves the cursor to the end. The **backspace** and **delete** keys can be used to remove characters from a command line. The **insert** key is used to initiate the insertion of characters. Pressing the insert key a second time exits the insert mode. If MATLAB recognizes an error after you have pressed ENTER, then MATLAB responds with a beep and a message that may help define the error. You can recall the command line using the up arrow key in order to edit the line. (If you make an error and are unable to determine its cause, ask your instructor for help.)

- Continuing Commands.

  MATLAB commands that do not fit on a single line can be continued to the next line using an ellipsis, which is three consecutive periods followed by ENTER.

- Stopping a Command.

  To stop execution of a MATLAB command press **Ctrl** and **C** simultaneously, then press ENTER. Sometimes this sequence must be repeated.

- Quitting.

  To quit MATLAB type **exit** or **quit** followed by ENTER.

Table of Contents

LABS

Lab 1. Matrices in MATLAB

Lab 2. Linear Systems

Lab 3. Matrix Operations

Lab 4. Homogeneous Systems, Echelon Forms, and Inverses

Lab 5. A Vector Space Example

Lab 6. Linear Combinations

Lab 7. Coordinates and Change of Basis

Lab 8. The Determinant Function

Lab 9. Inner Product Spaces

Lab 10. Orthogonal Sets

Lab 11. Plane Linear Transformations

Lab 12. Linear Transformations

Lab 13. The Eigenproblem

PROJECTS

APPENDICES

New Features in the Third Edition

- New sections have been added on
    - ▷ Vector space properties.
    - ▷ Translations using homogeneous coordinates.
- New and revised exercises.
- New instructional activities to engage students in hands-on investigations through experimentation and discovery.
- Revisions to time sensitive data.
- Improvements in clarity of writing throughout.

Topical Organization

The list below gives the sections that should be covered to incorporate the topics recommended by the Linear Algebra Curriculum Study Group (LACSG) for a one-semester course. The order of the topics reflects that recommended by LACSG; it also attests to the flexibility of these Labs.

Topic	Section	Subject
Matrices	1.1	Entering matrices into MATLAB
	3.1	Operations on matrices
Systems of equations	2.1	Reduction process
	4.1	Homogeneous systems
	4.2	Reduced row echelon form
	4.3	Matrix inverses
	Project 3.2	Least squares
Fundamental ideas	6.1	Linear combinations
	6.2	Span
	6.3	Independence/Dependence
	6.4	Basis and dimension
	7.2	Coordinates
Linear transformations	11.1	Graphics Experiments
	11.2	More Graphics Experiments
	12.1	Linear transformations
	12.2	Matrix of a linear transformation
	12.3	Range
	12.4	Kernel
Orthogonality	9.1	Inner product
	9.2	Norm
	9.3	Angles
	10.1	Orthonormal bases
	10.2	Projections
	10.3	Gram-Schmidt
Determinants	8.1	Determinants
Eigen problem	13.2	Eigenvalues and eigenvectors

Instructional Routines: Overview [4]

- **DOCTOR** Loads data for doctorates awarded annually in mathematics to U.S. citizens. (Least Squares Models)

- **EVECSRCH** Searches for eigenvectors of a 2×2 matrix using vectors around the unit circle. (Lab 13)

- **GSCHMIDT** Performs the Gram-Schmidt process on the columns of a matrix. (Lab 10)

- **HIGHJUMP** Loads men's high jump Olympic data. (Least Squares Models)

- **HOMSOLN** Produces a set of basis vectors for the solution set of a homogeneous system of equations. (Lab 12)

- **IGRAPH** Creates a graph associated with an incidence matrix. (Introduction to Graph Theory)

- **INVERT** Computes the inverse of matrix A using the row reduced echelon form of [A I]. (Lab 4)

- **LINCOMBO**† Interactive graphical activity to express one vector as a linear combination of two others. (Lab 7)

- **LISUB** Determines a linearly independent subset of a given set of vectors. (Lab 12)

- **LONGJUMP** Loads men's long jump Olympic data. (Least Squares Models)

- **LSQGAME**† An interactive graphical game to determine a least squares line. (Least Squares Models)

- **LSQLINE**† Constructs the equation and graph of the least squares line with evaluation and deviation display options. (Least Squares Models)

[4] $\dagger$ indicates that the routine employs the graphical user interface. File **alldesc.txt** contains the collection of help files for the instructional files listed. It can be printed to have an easy reference document.

- **M1500RUN** Loads men's 1500 meter run Olympic data. (Least Squares Models)
- **MAPCIRC** Displays the image of the unit circle mapped by 2×2 matrix. (Lab 13)
- **MATDAT1** Loads a data set for practice with matrix operations. (Lab 3)
- **MATDAT2** Loads an alternate data set for practice with matrix operations. (Lab 3)
- **MATOPS** Generates a screen summarizing matrix operations. (Lab 3)
- **MATRIXMAPS**† Linear transformations and translations graphically. (Lab 11)
- **MATVEC**† Interactive exploration of function $F(x) = A * x$, where A is a 2×2 matrix. (Lab 13)
- **MODN** Generates the remainder of x divided by n. (Secret Codes)
- **PLANELT** Displays graphical results of plane linear transformations. (Lab 11)
- **PROJECT** Graphically shows the projection of one vector onto another; options for 2D and 3D. (Lab 10)
- **PROJXY** Graphically shows the projection of a 3D vector onto the xy-plane using constructive stages. (Lab 10)
- **RATIONAL** Displays the rational form of a matrix. (Lab 3)
- **REDUCE** Performs row operations under user control using menu selection. (Lab 2)
- **ROWECH** Instructional aid for teaching reduced row echelon form; provides hints, help, and checking. (Lab 4)
- **ROWOP**† Performs row operations under user control via mouse interaction. (Lab 2)
- **RREFQUIK** Provides a quick movie form of reduction to reduced row echelon form. (Lab 4)
- **RREFSTEP** Provides a step-by-step annotated form of reduction to reduced row echelon form. (Lab 4)

- **RREFVIEW** Provides a slow movie form of reduction to reduced row echelon form that pauses after each step. (Lab 4)

- **SYMROWOP**† Row reduction of a symbolic matrix; tracks potential divisions by zero. Requires Symbolic Math Toolbox. (Lab 2)

- **SYMRREF** Forms the reduced row echelon form of a symbolic matrix and records restrictions to avoid division by zero. Requires Symbolic Math Toolbox. (Lab 4)

- **UBALL** Graphical demonstration of the shape of unit balls in 2-space or 3-space for various norms. (Lab 9)

- **VAULTLSQ** Displays piecewise linear least squares model for men's Olympic pole vault event. (Least Squares Models)

- **VIZROWOP**† Provides a graphics visualization of row operations on 2×2 linear systems. (Lab 2)

- **W100DASH** Loads women's 100 meter dash Olympic data. (Least Squares Models)

- **W100FREE** Loads women's 100 meter freestyle swimming Olympic data. (Least Squares Models)

The instructional M-files are available from the authors or through the Companion Web Site at http://www.prenhall.com

LAB 1

Matrices in MATLAB

Topics: Basic MATLAB concepts: defining a matrix; = operator; matrix names; displaying entries, rows, and columns; suppressing display; the colon (:) operator. Complex numbers, complex conjugate.

Introduction

A matrix is a rectangular array used to store and organize data. The simple, versatile matrix is essentially the only kind of object with which MATLAB works. (The name MATLAB is an abbreviation for MATrix LABoratory.)

Section 1.1 shows how to enter data, matrices, into MATLAB , identify them with names, and how to address their entries. We also show how to view the contents of a matrix, and how to change individual entries, rows, or columns. The assignment operator = and the colon (:) command are discussed. These basic commands are used throughout this book.

Section 1.2 shows how to enter complex numbers and complex matrices. MATLAB does complex arithmetic automatically.

The exercises provide an opportunity to explore related ideas and gain familiarity with MATLAB .

Section 1.1

Getting DATA into MATLAB

To enter a matrix

$$\begin{bmatrix} 5 & -4 & 0 \\ -7 & 1 & 12 \\ 3 & 2 & 6 \end{bmatrix}$$

into MATLAB type the following:

$$[5 \ \ -4 \ \ 0 \ \ ; \ \ -7 \ \ 1 \ \ 12 \ \ ; \ \ 3 \ \ 2 \ \ 6]$$

Note:

- The matrix is enclosed in square brackets.
- Entries are separated by a space or by commas.
- Rows are separated by a semicolon. (The spaces beside the semicolons are optional and are used here for legibility.)

The display generated is

```
ans =

     5    -4     0
    -7     1    12
     3     2     6
```

Notice that no brackets are displayed and that MATLAB has assigned this matrix to the name **ans**.

> Every matrix in MATLAB must have a name. If you do not assign a matrix a name, then MATLAB assigns it to **ans**, which we call the default variable name.

To assign a matrix name we use the assignment operator =. For example

$$\boldsymbol{A} = [1 \;\; 2 \;\; 3 \;\; ; \;\; 4 \;\; 5 \;\; 6]$$

is displayed as

```
A =
     1     2     3
     4     5     6
```

Warnings:

- All rows must have the same number of entries.
- MATLAB distinguishes between upper and lower case letters. So matrix $\boldsymbol{B}$ is not the same as matrix $\boldsymbol{b}$.
- A matrix name can be reused. In such a case the 'old' contents are lost.

To assign a matrix but **suppress the display of its entries** follow the closing square bracket,], with a semicolon. The MATLAB command

$$\mathbf{A} \;=\; [1 \;\; 2 \;\; 3; \;\; 4 \;\; 5 \;\; 6];$$

assigns the same matrix to name $\boldsymbol{A}$ as above, but no display will appear. You can enter this command without retyping by using MATLAB 's edit feature. Press the **up arrow** key to display the previous command. Now just type the semicolon. (For more information on the edit features see the Introduction to MATLAB and Some of Its Features in the Preface.)

> To assign a currently defined matrix to a new name use the assignment operator =.

Command $\boldsymbol{Z} = \boldsymbol{A}$ assigns the contents of matrix $\boldsymbol{A}$ to matrix $\boldsymbol{Z}$. Matrix $\boldsymbol{A}$ is still defined.

To change an entry type the matrix name, the entry's location, the equals sign = , and the new value. For example

$$\boldsymbol{A}(2,1) = -12$$

sets the $(2,1)$ entry of matrix $\boldsymbol{A}$ to -12. The entire matrix with this new entry is displayed.

Seeing a Matrix

To see all of the contents of a matrix type its name. If the matrix is large the display may be broken into subsets of columns which are successively shown. For example type the command

hilb(9)

The display shows the first 7 columns followed by columns 8 and 9. (For information on command **hilb** type **help hilb**.) If a matrix is quite large the screen display will scroll too fast for you to see the matrix. You can drag the scroll bar to reveal portions of a display as needed. Alternatively type command **more on** followed by the matrix name or a command to generate it. Press the Space Bar to reveal more of the matrix. Continue pressing the Space Bar until the message '– more–' no longer appears near the bottom of the screen. Try this with **hilb(20)**. To disable this paging feature type command **more off**. (Use **help** for additional information on the **more** command.)

We have the following conventions to see a portion of a matrix in MATLAB . For purposes of illustration, type **A = hilb(5)**.

- To see the $(2,3)$ entry of $\boldsymbol{A}$ type

  A(2, 3)

- To see the 4th row of $\boldsymbol{A}$ type

  A(4, :)

- To see the first column of $\boldsymbol{A}$ type

  A(:, 1)

In the above situations the colon, **:**, is interpreted to mean 'all'. The colon can also be used to represent a range of rows or columns. For example typing

2 : 8

displays

```
ans =

     2     3     4     5     6     7     8
```

The step value or increment in command $2:8$ is understood to be 1. To increment by 3, we use command $2:3:8$. Try it. In general, the increment need not be a whole number. Try $1:.25:4$. Also try $2:-.3:-2.4$.

We can use the colon operator to display a subset of rows or columns of a matrix. As an illustration, to display rows 3 through 5 of matrix $\boldsymbol{A}$ type

$$\mathbf{A}(3:5,:)$$

Similarly columns 1 through 3 are displayed by typing

$$\mathbf{A}(:,1:3)$$

For more information on the use of the colon operator type **help colon**. The colon operator is very versatile in MATLAB but we will not need to use all of its features.

Exercises 1.1

Enter matrices $\boldsymbol{A}$, $\boldsymbol{B}$, and $\boldsymbol{C}$ into MATLAB .

$$\boldsymbol{A}=\begin{bmatrix} 4 & -3 \\ 2 & 1 \\ 0 & 6 \end{bmatrix} \qquad \boldsymbol{B}=\begin{bmatrix} 1 & 2 & 4 \\ 2 & 4 & 1 \\ 0 & 1 & 5 \end{bmatrix} \qquad \boldsymbol{C}=\begin{bmatrix} 5 \\ 8 \\ 7 \end{bmatrix}$$

Exercises 1 and 2 refer to matrices $\boldsymbol{A}$, $\boldsymbol{B}$, and $\boldsymbol{C}$.

1. On the line provided enter the command that performs the indicated action. Execute it in MATLAB .

a) Display all of $\boldsymbol{A}$. ______________________

b) Display only the second row of $\boldsymbol{A}$. ______________________

c) Display only the (3,2)-entry of $\boldsymbol{A}$. ______________________

d) Display only column 3 of $\boldsymbol{B}$. ______________________

e) Display the first two columns of $\boldsymbol{B}$. ______________________

f) Display the last 2 rows of $\boldsymbol{A}$. ______________________

2. Define a new matrix $\boldsymbol{D}$ having the same contents as $\boldsymbol{A}$ by typing the MATLAB command **D = A**. On the line provided enter the command that performs the indicated action where appropriate.

a) Make the (1,1)-entry of $\boldsymbol{D}$ equal to 12. ________________

b) Make the (3,2)-entry of $\boldsymbol{D}$ equal to -8. ________________

c) Type the command **E** = [**D C**]. Describe the contents of $\boldsymbol{E}$ in terms of $\boldsymbol{D}$ and $\boldsymbol{C}$.

____

d) Type the command **F** = [**D B**]. Describe the contents of $\boldsymbol{F}$ in terms of $\boldsymbol{D}$ and $\boldsymbol{B}$.

____

e) Type the command **G** = [**E;B**]. Describe the contents of $\boldsymbol{G}$ in terms of $\boldsymbol{E}$ and $\boldsymbol{B}$.

____

3. To enter a column matrix into MATLAB type its entries separated by semicolons as in

$$[1; 2; 3]$$

Perform the following in MATLAB .

a) Construct a column **c1** with entries $0, -1, 3, 5$.

b) Construct a column **c2** with entries $4, -2, 0, 7$.

c) Construct a matrix $\boldsymbol{H}$ whose columns are **c1** and **c2** without retyping any entries. Record the command you used below.

____

d) Construct a matrix $\boldsymbol{K}$ whose first two columns are both **c1** and whose third column is **c2**, without retyping any entries. Record the command you used below.

____

4. To enter a row into MATLAB type its entries separated by spaces as in

$$[1 \quad 2 \quad 3]$$

Perform the following in MATLAB .

a) Construct a row $\boldsymbol{r1}$ with entries $2, -1, 5$.

b) Construct a row $\boldsymbol{r2}$ with entries $7, 9, -3$.

c) Construct a matrix $\boldsymbol{M}$ whose rows are $\boldsymbol{r1}$ and $\boldsymbol{r2}$ without retyping any entries. Record the command you used below.

____

d) Describe the result of command 3∗**r1**.

____

e) Describe the result of command **r1** + **r2**.

____

f) Describe the result of command [**r1**; **r1** − **r2**; **r2**].

In Exercises 5 – 7 set up the system of equations that serves as a **mathematical model** for the problem statement. The system will have the form

$$\square\, x_1 + \square\, x_2 = \square$$
$$\square\, x_1 + \square\, x_2 = \square$$

where you replace the boxes by numerical values obtained from the problem statement.

5. Two amounts of money x_1 and x_2 total \$600. The amount x_1 is twice the amount x_2. (Solve the system.)

6. Let x_1 = the number of cupcakes and x_2 = the number of cookies a baker is to make in one hour. On the average it takes a baker 4 seconds to prepare a cupcake and 10 seconds to prepare a cookie. The selling price of a cupcake is \$.35 and that of a cookie is \$.25. If the total revenue from the baker's work in 1 hour is to be \$127.50, how many cupcakes and cookies must be made? (Solve the system.)

7. (Follow the ideas in Exercises 5 and 6, only now there will be three unknowns.) A length of pipe 50ft. long is to be cut into three sections of lengths x_1, x_2, x_3. It is required that length x_3 be equal to the sum of lengths x_1 and x_2 and that length x_3 be five feet longer than twice the length x_2. (Solve the system.)

Section 1.2

Complex Numbers and Matrices

MATLAB is designed to work with complex numbers and perform operations on them. A complex number z has the form

$$z = a + bi$$

where a and b are real numbers called the *real part* and the *imaginary part*, respectively. (If $b = 0$, then z is a real number.) The symbol i is called the *complex unit* and

$$i = \sqrt{-1}$$

To enter complex numbers into MATLAB we must first ensure that the complex unit is defined. When MATLAB first starts up, variables i (and j) are loaded with $\sqrt{-1}$. However, since the names i and j are commonly used for subscripts and indices, it is recommended that the command

i = sqrt(−1)

be entered before you begin assigning complex values to matrix elements. The preceding command displays

```
i =

        0 + 1.0000i
```

To assign a complex number like $7 - 3i$ to z type

$$z = 7 - 3 * i \quad \text{or} \quad z = 7 - 3i$$

The multiplication symbol $*$ is optional between the imaginary part -3 and the complex unit **i**.

Do not put spaces between the real part, the sign, the imaginary part, the $*$, and the complex unit.

The display generated by the preceding command is

```
z =

    7.0000 - 3.0000i
```

To enter a matrix with complex entries use the same form as with real numbers and enter the complex entries as above. (Note: the complex unit can be assigned any name and that name used to form complex entries. However, MATLAB always displays **i** for the complex unit.)

Exercises 1.2

Enter the following matrices into MATLAB :

C= [1 + 5i 3 2 − i ; 0 4 − i 6i]   **D**= [7 + 2i ; 4 + i]

1. What is the display after you entered matrix C? Record it here:

2. Type each of the following commands and record the result next to it. (Try to predict the result beforehand.)

a) C(1, 3) ________ **b)** C(1, :) ________ **c)** C(:, 2) ________ **d)** C(1, 2 : 3) ________

3. a) What matrix results after typing the command **[C D]**? Record the result here:

$$\begin{bmatrix} \quad & \quad \\ \quad & \quad \end{bmatrix}$$

b) Type the command [**D C**]. Is this the same as in part **a**)? Circle one: Yes No.

4. The *conjugate* of a complex number $a + bi$ is $a - bi$. Type **help conj** and then use **conj** to compute the conjugate of the matrix C and the conjugate of the matrix D. Record the results below.

5. Type **help real**. Use the **real** command to display the real parts of matrices C and D. Record the results below.

6. Type **help imag**. Use the **imag** command to display the imaginary parts of matrices C and D. Record the results below.

7. Let A be a matrix with all real entries, say, **A** = **hilb(5)**. Write a short description of the result of each of the following commands and explain why your result is correct.

a) **conj(A)** ____

b) **real(A)** ____

c) **imag(A)** ____

LAB 2

Linear Systems

Topics: Row operations for solving linear systems; routine **rowop**; routine **vizrowop**; routine **symrowop**; transpose operator; conjugate transpose; **size**; **who** and **whos**; **clear**.

Introduction

The 'Rule of Three' emerged in the 1990s as the paragon for mathematical instruction. The rule states that, where possible, all topics should be considered from numerical, graphical, and symbolical viewpoints. This lab demonstrates the Rule of Three for solving linear systems of equations. Subsequent labs illustrate the rule in other settings. This lab expressly assumes Section 1.1, namely the method for entering matrices into MATLAB .

Tacitly this lab assumes that linear systems in two or three variables have already been solved by hand. Section 2.1 presents the first step in automating the process. Most students are overjoyed to be relieved of the arithmetical precision that the reduction process requires. The numerical capability demonstrated by the routine **rowop** in Section 2.1 is augmented by the routine **vizrowop** in Section 2.2. **Vizrowop** graphically displays the reduction process for 2 by 2 systems; the exercises suggest extensions to more general systems. Section 2.3 presents a symbolic approach to solving linear systems based on the routine **symrowop**. Here the coefficients are permitted to be parameters as well as numerical constants.

The final steps in the process of using MATLAB to solve systems of equations by elimination are carried out in Lab 4. It is possible to proceed directly to that lab after this one has been completed.

Section 2.1

Row Operations Using Routine ROWOP

Given a linear system $\boldsymbol{CX} = \boldsymbol{B}$ we enter the coefficient matrix $\boldsymbol{C}$ and the right hand side $\boldsymbol{B}$ into MATLAB . Then we form the augmented matrix $\boldsymbol{A}$ in MATLAB by typing

A = [C B]

(Note: There is a space between **C** and **B**.) Now we are ready to apply row operations to the augmented matrix to obtain a linear system that is simpler to solve. To aid in this process we use the routine **rowop**, which permits us to concentrate on the strategy of the row reduction while MATLAB performs the arithmetic. For a description of **rowop** type **help rowop**. Read the display.

To use **rowop** to solve a linear system the general procedure is as follows:

- Type the command **rowop**.
- Enter the augmented matrix.

- Use the mouse to select row operations and to define data entries.
- Press ENTER after the last input item has been entered.

We apply this procedure to solve the linear system

$$\begin{aligned} 2x_1 - x_2 + x_3 &= 8 \\ x_1 + 2x_2 + 3x_3 &= 9 \\ 3x_1 - x_3 &= 3 \end{aligned}$$

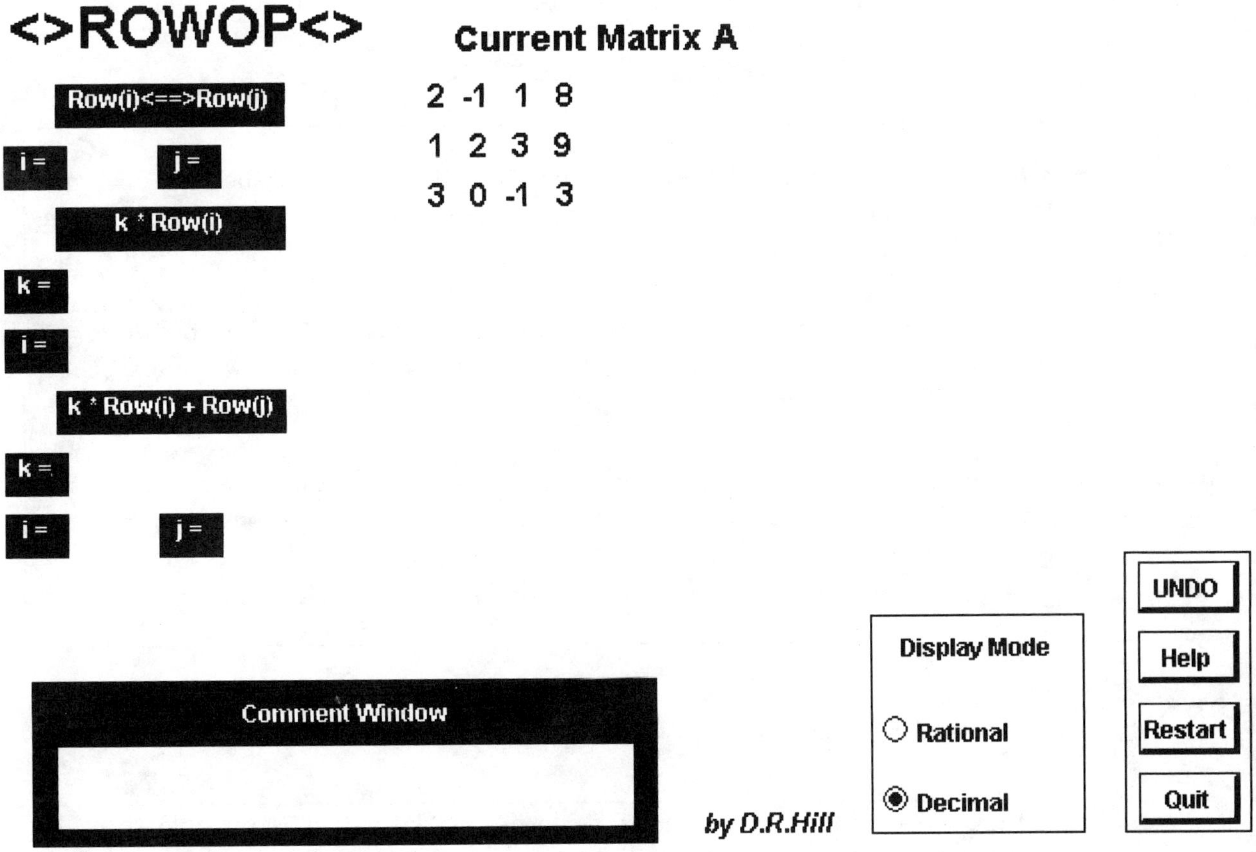

Figure 1.

Initiate the routine by typing **rowop**. The directions on the screen provide instructions for entering the augmented matrix. Here type [2 -1 1 8; 1 2 3 9; 3 0 -1 3] and press ENTER. This produces the **rowop** screen, which displays the Current Matrix A. See Figure 1. Verify that the entries on your screen are correct. (If any entry is incorrect, click the **Restart** button, press $\uparrow$ to retrieve the entered matrix, make any necessary changes, and press ENTER.)

We describe how to use the **rowop** screen to solve this system. The user chooses the proper strategy at each stage from the three elementary row operations, then MATLAB performs the arithmetic. The goal is to convert the augmented matrix $\boldsymbol{A}$ to a simpler form to see if the system of equations is consistent. If it is consistent, then it may be possible to *'eliminate'* more unknowns so that the solution can be read directly from the display. Box 1 describes the method for applying the three elementary row operations.

As a first step, for the present system we want to produce an equivalent system with entry A(1, 1) = 1. This can be achieved by the row interchange **Row(1)⇔Row(2)**. Following the instructions in Box 1, click in the box beside **i** = under **Row(i)⇔Row(j)**, type 1, click in the box beside **j** = , type 2, then press ENTER. **Rowop** performs the desired operation. A message displays the row operation that was used to produce the Current Matrix A. Check the matrix to verify that you have employed the correct strategy. If not, Box 2 describes how to correct errors.

- To switch two rows use **Row(i) ⇔ Row(j)**. Click in the box beside **i** =, type the row number, click in the box beside **j** =, type the other row number, then press ENTER.
- To multiply a row by a number use **k*Row(i)**. Click in the box beside **k** =, type the scalar multiple, click in the box beside **i** =, type the row number, then press ENTER.
- To multiply a row by a scalar, add it to a second row, and replace the second row by the sum, use **k*Row(i) + Row(j)**. Click in the box beside **k** =, type the multiplier, click in the box beside **i** =, type the row number, click in the box beside **j** =, type the other row number, then press ENTER.

Box 1.

If the Current Matrix $\boldsymbol{A}$ is not in the form you desired, it is possible to return to the previous matrix by clicking the UNDO button.

Box 2.

The next step is to produce an equivalent system with A(2, 1) = 0. This is achieved by the operation **k*Row(i) + Row(j)**, with k = -A(2, 1), i = 1, and j = 2. Follow the directions in Box 1 for entering these numbers in the boxes below **k*Row(i) + Row(j)** , then press ENTER. It is important to enter the value of k in the form k = -A(i, j), as we shall demonstrate shortly. Here, enter k = -A(2, 1), i = 1, and j = 2, then press ENTER. After performing this row operation, execute a similar row operation to produce A(3,1) = 0. Verify that the Current Matrix is

$$\begin{bmatrix} 1 & 2 & 3 & 9 \\ 0 & -5 & -5 & -10 \\ 0 & -6 & -10 & -24 \end{bmatrix}$$

To produce A(2, 2) = 1, use the elementary row operation **k*Row(i)** with k = 1/A(2, 2) and i = 2. After MATLAB performs the operation the Current Matrix is

$$\begin{bmatrix} 1 & 2 & 3 & 9 \\ 0 & 1 & 1 & 2 \\ 0 & -6 & -10 & -24 \end{bmatrix}$$

To produce A(3, 2) =0 use **k*Row(i) + Row(j)** with k = -A(3, 2), i = 2, and j = 3. The resulting matrix is

$$\begin{bmatrix} 1 & 2 & 3 & 9 \\ 0 & 1 & 1 & 2 \\ 0 & 0 & -4 & -12 \end{bmatrix}.$$

This shows that the system is consistent. (Why?) Finally, execute row operations to produce A(3, 3) = 1, A(2, 3) = 0, A(1, 3) = 0, and A(1, 2) = 0. The resulting matrix ***A*** is

$$\begin{bmatrix} 1 & 0 & 0 & 2 \\ 0 & 1 & 0 & -1 \\ 0 & 0 & 1 & 3 \end{bmatrix}$$

It follows that the solution to this system is $x_1 = 2, x_2 = -1, x_3 = 3$. Hence, this is the solution to the original system of equations too. (Why?)

It is important that the value of the multiplier k be entered in the form k = A(i, j), even though it is tempting to use the number displayed on the screen. To illustrate this point, let's solve the linear system

$$\begin{aligned} (1/3)x + (1/4)y &= 13/6 \\ (1/7)x + (1/9)y &= 59/63 \end{aligned}$$

Click on the Restart button. Define the augmented matrix ***A*** as

[1/3 1/4 13/6; 1/7 1/9 59/63]

Perform the row operation **k*Row(i)** with k = 3 and i = 1. Instead of using matrix notation, perform **k*Row(i) + Row(j)** with k = -0.1429, i = 1, and j = 2. Notice that the goal of producing A(2, 1) = 0 was not achieved because A(2, 1) = -0.00004286. The reason for the discrepancy is that we used the screen display's value of 0.1429 instead of the exact value of 1/7.

Click the UNDO button to return to the previous system. Then perform **k*Row(i) + Row(j)** with k = -A(2, 1), i = 1, and j = 2. The Current Matrix should now read

$$\begin{bmatrix} 1 & 0.75 & 6.5 \\ 0 & 0.003968 & 0.007937 \end{bmatrix}$$

Continue to solve this system using matrix notation. The solution is $x_1 = 5, x_2 = 2$.

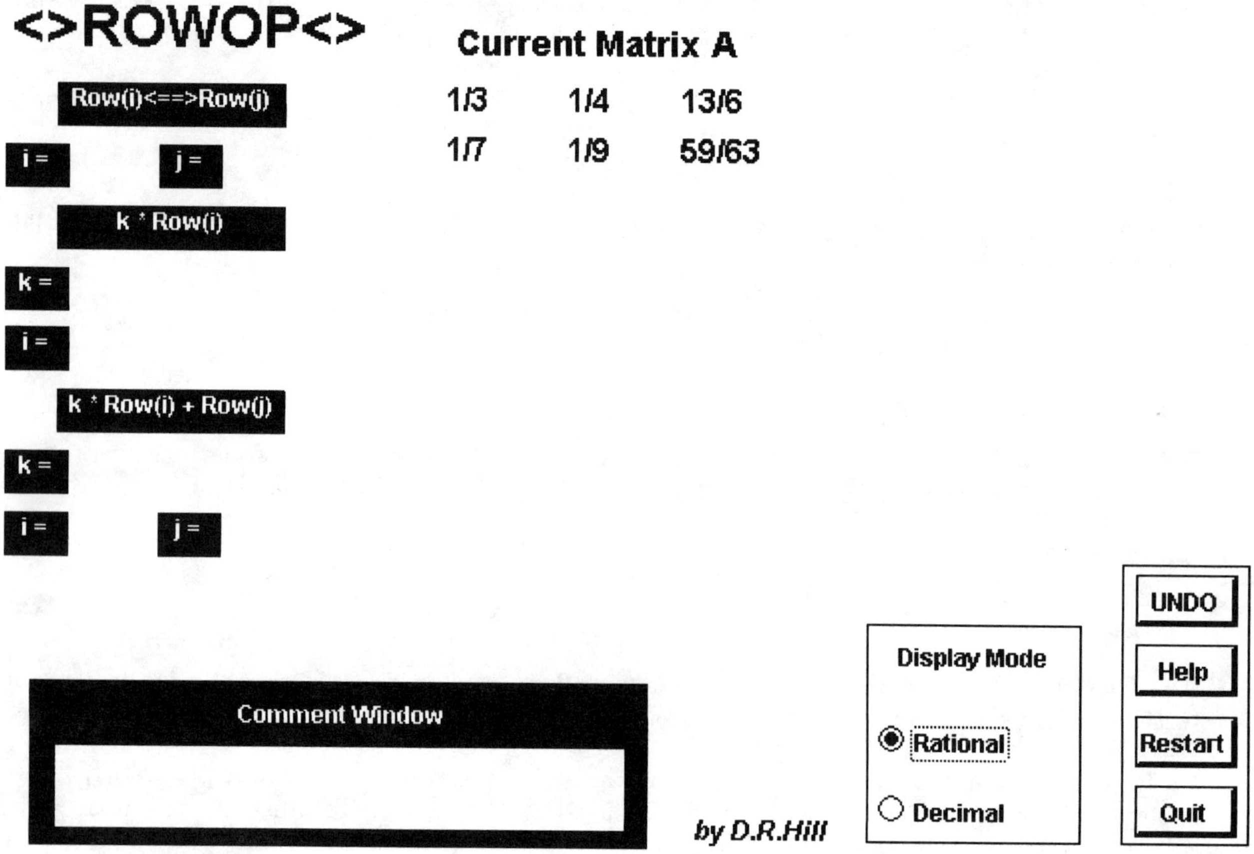

Figure 2.

An alternate approach makes use of the rational display mode. The default representation is decimal. Click the **Restart** button, then press the ↑ key to retrieve the matrix $\boldsymbol{A}$. (If necessary, you can retype the matrix.) In the **rowop** screen, click the **Rational** button under 'Display Mode.' The Current Matrix reflects the change, and it is possible to toggle between the two modes this way. See Figure 2. This system can be solved using the following sequence of elementary row operations.

- 3∗ Row(1)
- (−1/7)∗ Row(1) + Row(2)

- 252* Row(2)

- (−3/4)* Row(2) + Row(1)

Once again, the Current Matrix A yields the solution $x_1 = 5, x_2 = 2$.

This completes **rowop**. We have shown how to use all three elementary operations for solving linear systems that have a unique solution. The same kind of approach can be used for inconsistent systems and systems having infinitely many solutions. To exit **rowop** click the **Quit** button, then press ENTER to get the MATLAB prompt.

Exercises 2.1

In Exercises 1-4 use command **rowop** to solve the following systems of linear equations. Record your answer below the system.

1. $$\begin{aligned} 2x_1 + x_2 &= 3 \\ 3x_1 + 4x_2 &= 7 \end{aligned}$$

$x_1 =$ $x_2 =$

2. $$\begin{aligned} 2x_1 - 3x_2 &= 1 \\ x_1 - 2x_2 &= 3 \end{aligned}$$

$x_1 =$ $x_2 =$

3. $$\begin{aligned} 2x_1 + 3x_2 - x_3 &= 9 \\ -2x_1 + 3x_3 &= -5 \\ x_1 - x_2 + 2x_3 &= -3 \end{aligned}$$

$x_1 =$ $x_2 =$ $x_3 =$

4. $$\begin{aligned} 2x_1 - 3x_2 + 4x_3 + 2x_4 &= 12 \\ 3x_1 - 4x_2 - 2x_3 + x_4 &= 1 \\ 4x_1 + 5x_2 + x_3 - 6x_4 &= 21 \\ x_1 - x_2 - 2x_3 - x_4 &= 1 \end{aligned}$$

$x_1 =$ $x_2 =$ $x_3 =$ $x_4 =$

Problems 5-7 introduce the **transpose** of a matrix. Most textbooks denote the transpose of a real matrix $\boldsymbol{A}$ by either $\boldsymbol{A^t}$ or $\boldsymbol{A^T}$. However in MATLAB we use $\boldsymbol{A'}$.

5. In MATLAB type **A** = [1 2 3; 4 5 6] . What matrix is displayed when you type command **A**′?

Record the result here:

$$\begin{bmatrix} & & \\ & & \\ & & \end{bmatrix}$$

6. Let $\boldsymbol{A}$ be the matrix in Exercise 5.

a) Type **size(A)**. Type **size(A′)**.
How is the size of $\boldsymbol{A}$ related to the size of $\boldsymbol{A}'$?______________________________________

b) How are the rows of $\boldsymbol{A}$ related to the columns of $\boldsymbol{A}'$?______________________________________

7. How is the matrix $(\mathbf{A}')'$ related to the matrix $\mathbf{A}$?______________________________________

Exercises 8 and 9 deal with complex numbers in MATLAB . Complex numbers are discussed in Section 1.2 and that material should be referred to as needed. Enter the following matrix into MATLAB :

$$\mathbf{C} = \ [1+i \quad 3 \quad 2-i \quad ; \quad 0 \quad 4-i \quad i]$$

8. Describe in words the result of command $\mathbf{C}'$. (Hint: See Exercise 4 in Section 1.2.)

____

9. On a complex matrix the result of the prime (′) operator in MATLAB is called the *conjugate transpose.*

a) If the matrix is real, its conjugate transpose is the same as the ________________.

b) Enter the following matrix into MATLAB :

$$\mathbf{Q} = \ [1-2i \quad 3i \quad 2+i \quad ; \quad 0 \quad 4-3i \quad i \quad ; \quad -5+i \quad 6-2i \quad 3-i]$$

Compute $\boldsymbol{M} = \boldsymbol{Q} + \boldsymbol{Q}'$ and record it.

$$\boldsymbol{M} = \begin{bmatrix} & & \\ & & \\ & & \end{bmatrix}$$

Compute $\boldsymbol{P} = \boldsymbol{Q} - \boldsymbol{Q}'$ and record it.

$$\boldsymbol{P} = \begin{bmatrix} & & \\ & & \\ & & \end{bmatrix}$$

Show that $\boldsymbol{M} = \boldsymbol{M}'$. (A complex matrix equal to its conjugate transpose is called *Hermitian.*)

Show that $\boldsymbol{P} = -\boldsymbol{P}'$. (A complex matrix equal to the negative of its conjugate transpose is called *skew Hermitian.*)

Show that $\boldsymbol{Q} = \frac{1}{2}(\boldsymbol{M} + \boldsymbol{P})$.
Choose several other square complex matrices $\boldsymbol{Q}$ and verify that the preceding properties are still valid. If $\boldsymbol{Q}$ is a real square matrix are these properties still valid? Explain.

____

____

____

10. Use the command **rowop** to solve the following linear system with complex coefficients.

$$\begin{aligned} x_1 - ix_2 &= 0 \\ ix_1 - \ x_2 &= -2 \end{aligned}$$

Record the solution to the system: $x_1 =$ __________ $x_2 =$ __________

11. In MATLAB each matrix currently defined is assigned a name. To determine the names currently in use type command **who**. Also investigate command **whos**. Note that information about the type of matrix, real or complex, is indicated. To remove a variable from the current workspace, we use the **clear** command. Before using this command read the information in **help clear**. Try these commands with your current session of MATLAB .

Solving Math Models For each of the following word problems construct a linear system that models the relationships described and solve that system. Assign variable names to the quantities involved and use the problem data to develop equation relationships among the variables. Check to make sure that your solution makes sense within the context of the problem.

12. A gadget assembler currently works on two products, 'whirlies' and 'flingers'. It takes 2/3 of an hour to assemble a 'whirly' and 4/5 of an hour to assemble the more intricate 'flinger'. The components for each whirly cost \$4.90 and those for a flinger \$6.50. How many of each type of gadget can be made in 8 hours if the shop spends \$61.90 on the components required. (Hint: Let x_1 = the number of whirlies produce and x_2 = the number of flingers. Set up a time equation and then a cost equation.)

13. A small investment club has \$24,000 to invest in three stock plans, named A, B, and C. The club decides to invest twice as much in plan B as in plan C. The interest rates for each plan are respectively 10%, 8%, and 6% and the interest earned at the end of the year is required to be \$2000. How much should be invested in each plan?

14. A parabola $p(x) = ax^2 + bx + c$ is to be constructed through points $(1, 2)$, $(2, 4)$, and $(4, 14)$. Find the coefficients a, b, and c so that $p(1) = 2$, $p(2) = 4$, and $p(4) = 14$.

To generate a graph of the parabola you have constructed proceed as follows. In MATLAB enter the values of a, b, and c into a row matrix **p**. Then type commands

```
x=0:.1:5;
y=polyval(p,x);
figure,plot(x,y)
```

Section 2.2

Visualizing Row Operations

We have seen how **rowop** carries out the numerical details of elimination while permitting you to focus on the strategy of the row reduction process itself. Now we take this process one step further using the routine **vizrowop** to provide a geometric setting for the elementary row operations.[1] The exercises at the end of the section play a critical role in assessing whether you have achieved a firm understanding of the elimination process. Type **help vizrowop** for instructions. Then type **vizrowop** to initiate the routine. We illustrate the process by viewing what happens geometrically when the linear system

$$\begin{aligned} 3x + 4y &= 11 \\ x + 2y &= 5 \end{aligned}$$

is solved numerically. As prompted, enter the augmented matrix [3 4 11; 1 2 5]. Examine the coefficients on the **vizrowop** screen. (If necessary, click the **Restart** button to make any corrections.)

Vizrowop displays the graphs of the two lines. We are ready to begin the elimination process and to view how the geometry forms the underpinning for the algebra.

Box 3 summarizes the method for performing the three elementary row operations in **vizrowop**. The operations are similar to those in the routine **rowop**.

- To switch the two rows click the button **Row(1) ⇔ Row(2)**.
- To multiply a row by a number click the button **k∗Row(i)**, define k and i, then press ENTER.
- To multiply a row by a number, add it to a second row, and replace that second row by the sum, click the button **k∗Row(i) + Row(j)**, define k, i, and j, then press ENTER.

Box 3.

[1]The routine **vizrowop** is restricted to two linear equations in two unknowns.

The first step for this system is to produce A(1, 1) = 1. One way to proceed is to switch the rows, so click the button for **ROW(1) ⇔ ROW(2)**. This produces a new screen in which the rows are switched, so that the order of the equations is reversed. (See Figure 3.) Geometrically, the graphs in the Current System remain the same as the graphs in the Previous System because the equations have not changed at all. Look closer, however. Notice that the color and the texture of the lines change to reflect the color coding with the corresponding equations.

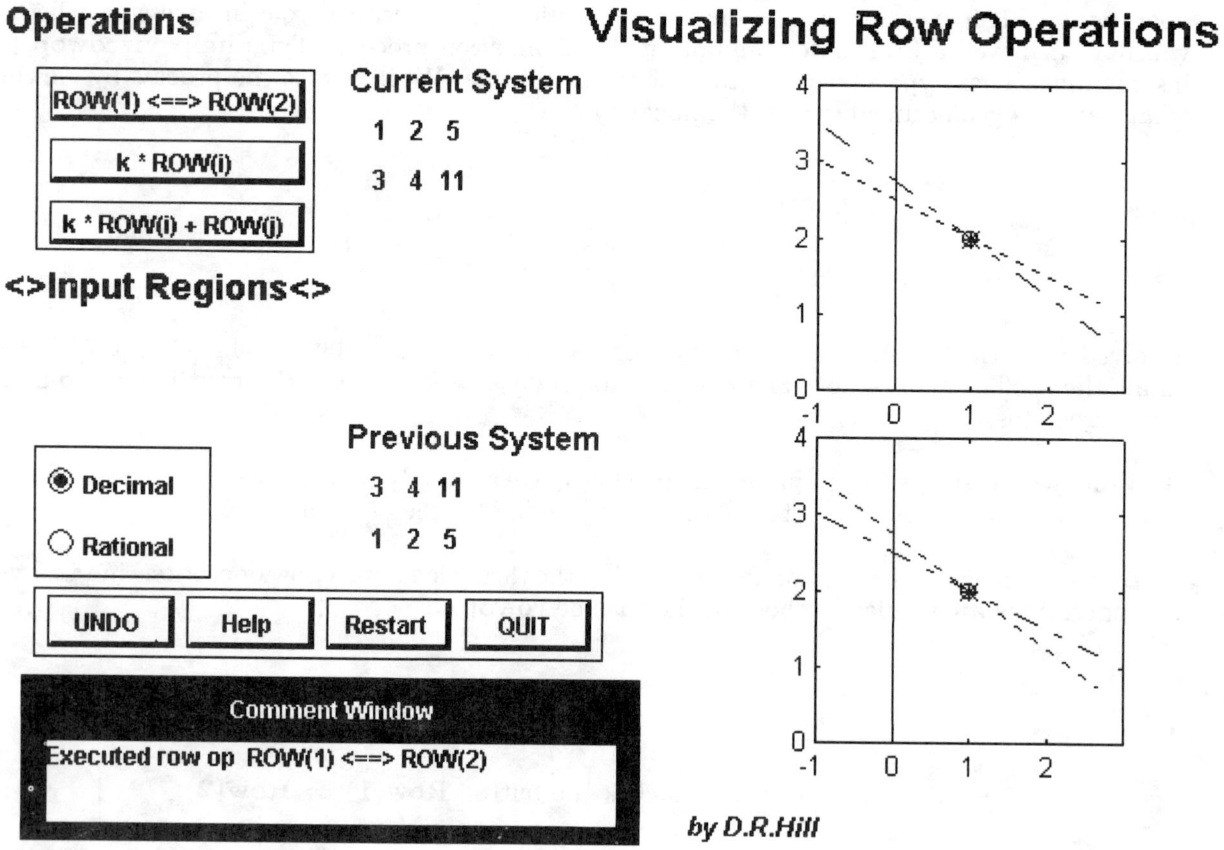

Figure 3.

The second step is to produce A(2, 1) = 0 in the matrix. This is achieved by the elementary row operation **k∗ROW(i) + ROW(j)**. Click on this button to open a section of the screen called Input Regions. As indicated in Box 3, click the box beside ***k*** =, then type -A(2, 1). Similarly, enter i = 1 and j = 2, then press ENTER. The matrix reflects this change. This row operation causes a crucial geometric change, one that lies at the heart of the elimination process: the graph of the second equation becomes a horizontal line. Notice that the two lines meet at the same point as the lines in the Previous System.

The next step, to produce A(2, 2) = 1 in the Current System, is achieved by the elementary row operation **k∗ROW(i)**. So click on this button, define k = 1/A(2, 2) and i = 2, then press ENTER. The resulting graphs remain the same, which is not surprising. (See Exercise 3.)

The final step is to produce A(1, 2) = 0 by the row operation **k∗ROW(i) + ROW(j)**. Predict how the graphs will change.

The graph of the first equation will ____

The graph of the second equation will ____

Now let **vizrowop** perform the numerical and graphical steps. The effect of this elementary row operation is to convert the graph of the first equation to a vertical line. Once again the point of intersection of the lines remains fixed. The screen now depicts the solution in two ways. Numerically the matrix yields $x = 1, y = 2$. Graphically the intersection of the horizontal line $y = 2$ and the vertical line $x = 1$ is the point $(1, 2)$.

Box 4 summarizes the geometric effects of each elementary row operation on a system of two equations in two unknowns (assuming the lines are not parallel). It also describes the overall process.

- When the order of the equations is reversed the graphs remain the same but their colors and texture change.
- When a row is multiplied by a scalar the graphs remain the same.
- When a row is multiplied by an appropriate scalar and added to a second row, the graph of the equation corresponding to the second row becomes either a vertical line or a horizontal line.
- Overall, the process of elimination converts a system of two lines that meet at a point to a system with one horizontal line and one vertical line that meet at the same point.

Box 4.

Not all linear systems of two equations in two unknowns meet at one point. The exercises consider the other possibilities. To exit from **vizrowop** click the **QUIT** button, then press ENTER for the MATLAB prompt.

Exercises 2.2

1. Explain what happens geometrically when each of the following row operations is performed on a system of three equations in three unknowns:

a) Two rows are interchanged.

b) A row is multiplied by a nonzero scalar.

c) A scalar multiple of one row is added to a second row, and the sum replaces the second row.

2. Explain what happens geometrically when a linear system of three equations in three unknowns is converted to an equivalent system in which as many variables as possible have been eliminated from the equations.

3. Explain why it is 'not surprising' that multiplying an equation by a nonzero scalar does not change the graphs of the system.

4. Use **vizrowop** to solve each system and to visualize the geometric process of elimination. Describe how the geometry reveals the number of solutions.

a)
$$\begin{aligned} 6x + 3y &= 1 \\ 6x + 9y &= 3 \end{aligned}$$

b)
$$\begin{aligned} 2x + 3y &= 1 \\ 6x + 9y &= 2 \end{aligned}$$

c)
$$\begin{aligned} 2x + 3y &= 1 \\ 6x + 9y &= 3 \end{aligned}$$

d)
$$\begin{aligned} 2x + 3y &= 0 \\ 6x + 9y &= 0 \end{aligned}$$

5. Use **vizrowop** to view the geometric process of elimination in solving the following system

$$\begin{aligned}(1/3)x + (1/4)y &= 13/6\\ (1/7)x + (1/9)y &= 59/63\end{aligned}$$

a) Use rational display.

b) Use decimal display.

Section 2.3

Symbolic Row Operations

MATLAB can perform symbolic manipulations, provided you have the Symbolic Math Toolbox. In this section we exploit such capabilities for solving systems of equations. The routine **symrowop**, an instructional m-file accompanying this manual, has been developed for this purpose. We first discuss **symrowop**, then provide an example to illustrate its use. This is followed by a sequence of exercises to further explore row operations.

To begin, type **help symrowop** for instructions regarding the use of the routine. Notice that **symrowop** is restricted to matrices of size at most 5×5 and that matrix notation cannot be used to define the value of k in row operations **k*Row(i) + Row(j)** or **k*Row(i)**.

In MATLAB symbolic matrices are matrices whose entries are symbolic expressions like $5 + b$, $\sin(t)$, $b - 3 * c$, $a/(b + c)$, or $c \wedge 3$. A simple type of symbolic matrix that we encounter later has all of its entries numerical values except one entry is a single letter like a. Symbolic matrices are entered much like numeric matrices: enclose the entries in square brackets, with commas separating entries and semicolons separating rows. However to inform MATLAB the matrix is symbolic we must enclose the square brackets in single quotes and precede this with the designation **sym**. For example, the command

$$\boldsymbol{A = \mathrm{sym}('[a, 5 * b, 1 - c; 0, c/d, f]')}$$

produces output

```
A =

[ a, 5*b, 1-c]
[ 0, c/d,   f]
```

If any matrix entries are expressions then the operation symbols $+, -, *, /$, and $\wedge$ must be explicitly indicated. A common mistake is to type $5b$ instead of $5 * b$; multiplication must be explicitly indicated. Parentheses are highly recommended so that the order of operations is clearly indicated. For instance $\frac{ab}{c}$ should be entered as $(a * b)/c$.

Now we apply **symrowop** to a specific problem. Consider the linear system

$$\begin{aligned} -2x + y &= 1 \\ cx + y &= 2 \end{aligned}$$

Geometrically, it is advantageous to write the equations of the system in slope-intercept form:

$$\begin{aligned} y &= 2x + 1 \\ y &= -cx + 2 \end{aligned}$$

The graph of the first equation is a line with slope $m = 2$ and y-intercept $b = 1$. The graph of the second equation is a line with slope $m = -c$ and y-intercept $b = 2$. Since the lines will meet whenever they are not parallel, the linear system is consistent provided $c \neq -2$. Example 1 uses **symrowop** to obtain this result symbolically, and then to extend it.

Example 1. Find all values of c for which the following linear system is consistent.

$$\begin{aligned} -2x + y &= 1 \\ cx + y &= 2 \end{aligned}$$

Find the solution of the system when it is consistent.

To initiate the routine type **symrowop**. Then enter the augmented matrix for the system by typing **sym('[−2, 1, 1; c, 1, 2]')**. The **symrowop** screen resembles the screen for **rowop**, so the use of the elementary row operations should be familiar. The first step is to produce A(1,1) = 1, so multiply Row(1) by $k = -1/2$. This yields

$$\begin{bmatrix} 1 & -1/2 & -1/2 \\ c & 1 & 2 \end{bmatrix}$$

The next step is to produce A(2, 1) = 0, which is achieved by the operation **k*Row(i) + Row(j)**, with k = -c, i = 1, and j = 2. Remember that you cannot use k = -A(2, 1). This yields

$$\begin{bmatrix} 1 & -1/2 & -1/2 \\ 0 & 1/2 * c + 1 & 1/2 * c + 2 \end{bmatrix}$$

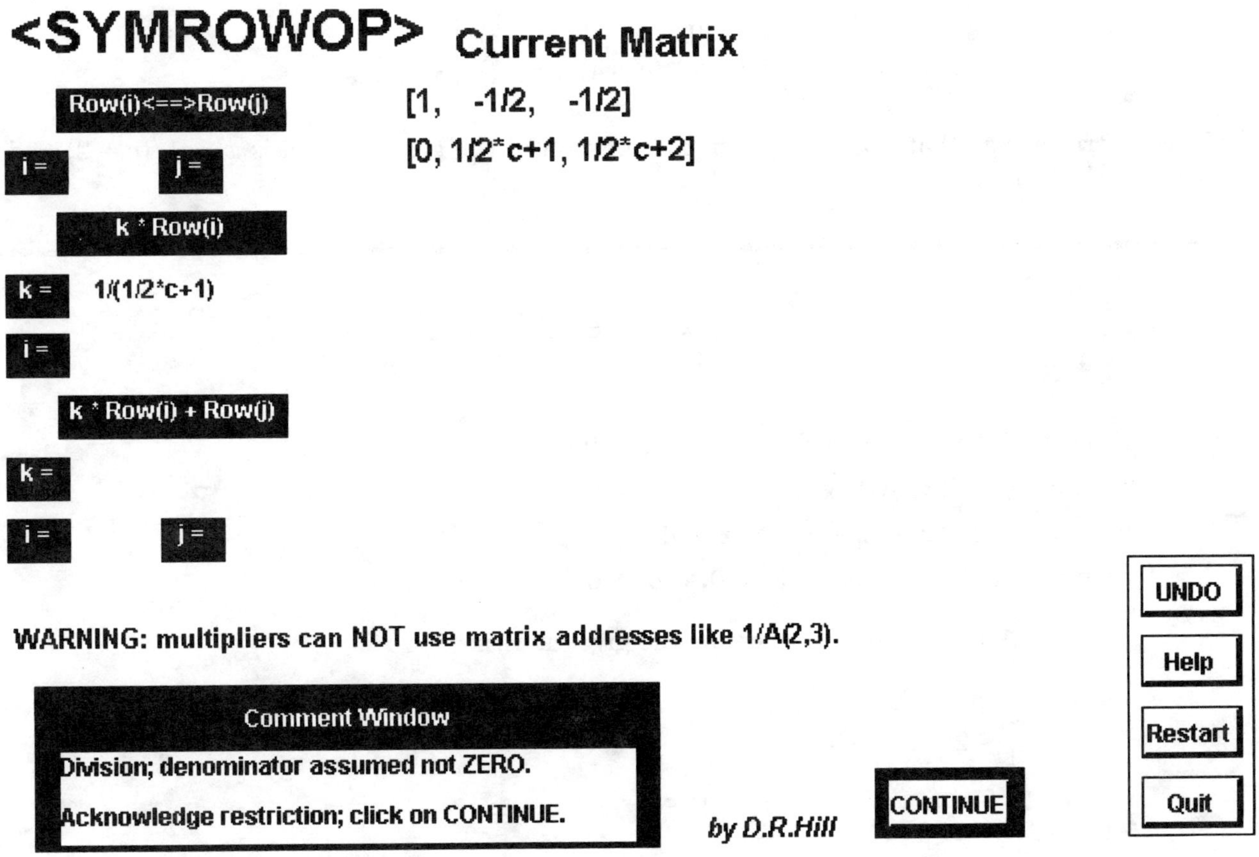

Figure 4.

See Figure 4. Now, multiply Row(2) by $k = 1/(1/2 * c + 1)$. As soon as you press the Tab key or click in the box beside **i =** routine **symrowop** detects that you are actually dividing the entries of the second row by $(1/2 * c + 1)$ and displays a message that this value is assumed to be different from zero. A confirmation of this requirement must be made by clicking the 'CONTINUE' button that appears. No row operation will be performed until you click on this button. (See Figure 4.) After you acknowledge this restriction, click in the box beside **i =** , enter 2, and press ENTER. MATLAB displays the matrix

$$\begin{bmatrix} 1 & -1/2 & -1/2 \\ 0 & 1 & (c+4)/(c+2) \end{bmatrix}$$

The restriction that $(1/2 * c + 1) \neq 0$ warns you that this row operation can be performed numerically only when $c \neq -2$. This result confirms the conclusion obtained by the geometric reasoning above. The symbolic capability allows you to extend the result by producing A(1, 2) = 0. This yields the reduced form

$$\begin{bmatrix} 1 & 0 & 1/(c+2) \\ 0 & 1 & (c+4)/(c+2) \end{bmatrix}$$

This matrix means that if $c \neq -2$, then the solution of the given system is $x = 1/(c+2)$ and $y = (c+4)/(c+2)$.

Exercises 2.3

1. What is the output of each MATLAB command?

a) $\boldsymbol{b = \text{sym}('[x; y; 0]')}$

b) $\boldsymbol{b = \text{sym}('[x-1, -2; -3, x-4]')}$

c) $\boldsymbol{b = \text{sym}('[1, 2, 3, 0; 4, 5, 6, 0; 7, 8, c, 0]')}$

2. Write a MATLAB command to define the given matrix.

a) $M = \begin{bmatrix} 1 & -3 & c \\ -2 & 6 & -5 \end{bmatrix}$

b) $C = \begin{bmatrix} 5-x & -8 & -1 \\ 4 & -7-x & -4 \\ 0 & 0 & 4-x \end{bmatrix}$

c) $\text{AUG} = \begin{bmatrix} 5-x & -8 & -1 & 0 \\ 4 & -7-x & -4 & 0 \\ 0 & 0 & 4-x & 0 \end{bmatrix}$

In Exercises 3 to 8 find all values of c for which the system is consistent. Find the solution of each consistent system.

3. $\begin{aligned} x + 2y &= 1 \\ 3x + cy &= 5 \end{aligned}$ $c =$__________ Solution(s): __________

4. $\begin{aligned} x - 2y &= 1 \\ 3x + cy &= -1 \end{aligned}$ $c =$__________ Solution(s): __________

5. $\begin{aligned} 3x + \ \ y &= 1 \\ cx + 4y &= 0 \end{aligned}$ $c =$__________ Solution(s): __________

6. $\begin{aligned} cx - y &= 1 \\ 3x + y &= -1 \end{aligned}$ $c =$__________ Solution(s): __________

7. $\begin{aligned} 3x + \ \ y &= 5 \\ 6x + 2y &= c \end{aligned}$ $c =$__________ Solution(s): __________

8. $\begin{aligned} 6x - 3y &= c \\ 4x - 2y &= 12 \end{aligned}$ $c =$__________ Solution(s): __________

9. Find all values of c for which the system (i) is inconsistent, (ii) has a unique solution, (iii) has infinitely many solutions.

a) $\begin{aligned} x + \qquad\quad y &= 2 \\ x - (c^2 - 5)y &= c \end{aligned}$ (i)__________ (ii) __________ (iii) __________

b) $\begin{aligned} x + \qquad\quad y &= -2 \\ x - (c^2 - 3)y &= c \end{aligned}$ (i)__________ (ii) __________ (iii) __________

10. Determine the value(s) of c such that the matrix is the augmented matrix of a consistent system.

a) $\begin{bmatrix} 1 & -3 & c \\ -2 & 6 & -5 \end{bmatrix}$ $c =$__________

b) $\begin{bmatrix} 1 & c & -2 \\ -2 & 1 & 5 \end{bmatrix}$ $c =$__________

11. Determine the value(s) of c such that $\boldsymbol{A}$ is the augmented matrix of a consistent linear system,

$$\boldsymbol{A} = \begin{bmatrix} 1 & 2 & 3 \\ 4 & 5 & 5 \\ 7 & 8 & c \end{bmatrix}$$ $c =$__________

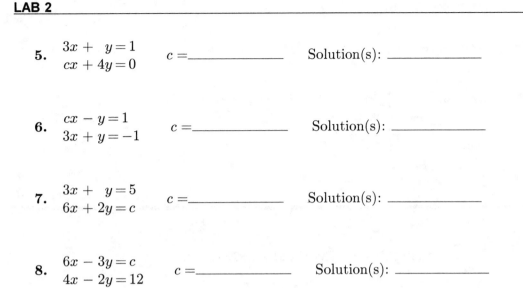

12. Determine the value(s) of c such that $\boldsymbol{A}$ is the augmented matrix of a consistent linear system,

$$\boldsymbol{A} = \begin{bmatrix} 1 & 2 & 3 \\ 4 & 5 & 5 \\ 8 & 7 & c \end{bmatrix} \qquad c = \underline{\hspace{4cm}}$$

<< NOTES; COMMENTS; IDEAS >>

LAB 3

Matrix Operations

Topics: Matrix operations in MATLAB $(+, -, *, ^\wedge)$; transpose of a matrix; commands **eye**, **ones**, **zeros**, **polyval**, **rand**, **randn**; display formats; an application on population growth.

Introduction

Here we introduce fundamental matrix operations in MATLAB together with several useful commands for generating special types of matrices.

Section 3.1 shows how to add, subtract, and multiply matrices in MATLAB . (There is no operation of division of matrices.) If the matrices involved are not compatible for the operation, then MATLAB will display an error message. We also explore how to multiply a matrix by a scalar, compute powers of a matrix, and take the transpose of a matrix.

Section 3.2 introduces MATLAB commands for generating special matrices. These commands are useful because we need not enter the individual elements of these special matrices. A number of exercises informally deal with ideas that play important roles later.

Section 3.3 discusses options for displaying matrices in decimal and fraction form.

Section 3.4 employs a model of population growth to illustrate the use of matrix multiplication for predicting long term behavior.

Section 3.1

Matrix Algebra

In MATLAB type

matops

You will see a display of the matrix operations available in MATLAB . In the case of addition and subtraction of matrices the form of the MATLAB operation is the same as in your text. However, there are variations in multiplication, scalar multiplication, exponentiation, and the transpose. Table 1 shows the book form and the MATLAB form.

The matrix sum $\boldsymbol{A}+\boldsymbol{B}$ and the matrix-vector product $\boldsymbol{A}*\boldsymbol{b}$ have geometric visualizations that enhance an understanding of the algebraic operations. We describe here three later sections that can be covered now in conjunction with the current section; they do not require any further prerequisites.

The matrix sum $\boldsymbol{A}+\boldsymbol{B}$ has a simple interpretation as the diagonal of a parallelogram when $\boldsymbol{A}$ and $\boldsymbol{B}$ are column matrices (or row matrices) with two entries. The routine **lincombo**, presented in Section 7.1, exploits this interpretation.

The matrix-vector product $\boldsymbol{A}*\boldsymbol{b}$ of an $m \times n$ matrix $\boldsymbol{A}$ and an $n \times 1$ vector $\boldsymbol{b}$ can be interpreted as a linear transformation from R^m to R^n. Section 12.1 explores this interpretation. A particular case of such a product is developed in Section 13.1, where the routine **matvec** views the product $\boldsymbol{A} * \boldsymbol{x}$ of a 2×2 matrix $\boldsymbol{A}$ times a 2×1 vector $\boldsymbol{x}$ as a linear transformation from a point $\boldsymbol{x}$ on the unit circle to another point $\boldsymbol{A} * \boldsymbol{x}$ on the unit circle.

	Book Form	MATLAB Form
matrix sum	$\boldsymbol{A} + \boldsymbol{B}$	$\boldsymbol{A} + \boldsymbol{B}$
matrix difference	$\boldsymbol{A} - \boldsymbol{B}$	$\boldsymbol{A} - \boldsymbol{B}$
matrix products	$\boldsymbol{AB}$	$\boldsymbol{A} * \boldsymbol{B}$
scalar multiple	$t\boldsymbol{A}$	$t * \boldsymbol{A}$
powers	$\boldsymbol{A}^k$	$\boldsymbol{A}$^$\boldsymbol{k}$
transpose	$\boldsymbol{A}^T$ or $\boldsymbol{A}^t$	$\boldsymbol{A}'$

Table 1.

MATLAB has a command to evaluate polynomials. Let $p(x) = 2x^3 + 4x - 7$, then define a row (or column) vector of coefficients of $p(x)$ as $[2\ 0\ 4\ -7]$. The polynomial must be arranged in decreasing powers of the variable and note that if a term is missing we assign a zero coefficient to its position. Example 1 shows how to use the MATLAB command **polyval** to compute the value of a polynomial at a given x-value and at a set of values. It also shows how MATLAB can be used to draw the graph of a polynomial.

Example 1. Compute $p(-1)$ for $p(x) = 2x^3 + 4x - 7$. Evaluate $p(x)$ for the set $\{-2, -1, 0, 1, 2\}$, and draw its graph in the interval $[-2, 2]$.

To calculate $p(-1)$, enter the following commands in MATLAB :

```
p=[2 0 4 -7]
x = -1
y = polyval(p,x)
```

Check by hand that $p(-1) = -13$. To evaluate $p(x)$ for the given set, enter the members of the set as a row matrix before applying the **polyval** command:

```
x = [-2 -1 0 1 2]
y = polyval(p,x)
```

Again check by hand that $p(-2) = -31$, $p(0) = -7$, $p(1) = -1$, and $p(2) = 17$. To graph $p(x)$ over $[-2, 2]$ enter the following commands: (The first command defines a vector of values starting with -2 in steps 0.1 to 2 which are used as x-values for the graph generated by the **plot** command.)

```
x= -2:0.1:2;
y = polyval(p,x);
poly(x,y)
```

Example 1 showed how to evaluate a polynomial at a row matrix corresponding to a set of x-values. We can evaluate a polynomial $p(x)$ at a square matrix of any size. For instance, for $p(x)$ in Example 1 and

$$\boldsymbol{A} = \begin{bmatrix} 1 & 2 & 3 \\ 0 & 0 & 4 \\ 0 & 0 & 0 \end{bmatrix}$$

to compute $p(\boldsymbol{A})$ we compute the expression $\mathbf{2A^3 + 4A - 7I}$ where

$$\boldsymbol{I} = \begin{bmatrix} \mathbf{1} & \mathbf{0} & \mathbf{0} \\ \mathbf{0} & \mathbf{1} & \mathbf{0} \\ \mathbf{0} & \mathbf{0} & \mathbf{1} \end{bmatrix}$$

In MATLAB we use the following commands:

```
A = [1 2 3; 0 0 4; 0 0 0]; I = eye(3);
2*A^3+4*A-7*I
```

which produces the result

$$\begin{bmatrix} \mathbf{-1} & \mathbf{12} & \mathbf{34} \\ \mathbf{0} & \mathbf{-7} & \mathbf{16} \\ \mathbf{0} & \mathbf{0} & \mathbf{-7} \end{bmatrix}$$

Verify this result by hand. It is instructive to point out that the command **polyval(p,A)** can not be used to evaluate $\mathbf{2A^3 + 4A - 7I}$. To demonstrate use vector **p** from Example 1 and the preceding matrix $\boldsymbol{A}$ in command

```
B=polyval(p,A)
```

This command produces $\boldsymbol{B} = \begin{bmatrix} \mathbf{-1} & \mathbf{17} & \mathbf{59} \\ \mathbf{-7} & \mathbf{-7} & \mathbf{137} \\ \mathbf{-7} & \mathbf{-7} & \mathbf{-7} \end{bmatrix}$. The command **polyval(p,A)** evaluates $\boldsymbol{p(x)}$ at each entry of **A**. For more information on this command, consult **help polyval**. See also the command **polyvalm**.

Exercises 3.1

1. To gain experience with matrix operations proceed as follows. Type

matdat1

This command loads a set of matrices so that we can perform MATLAB operations. You will see the matrices displayed on your screen. To see MATLAB display them type their names: **A**, **B**, **C**, **D**, **x**. If you forget what matrix names you have used type **who**. Command **who** displays the names of matrices currently in use. (The matrices in **matdat1** are shown here for ease of reference. An alternate set is available by using command **matdat2**).

With the matrices from routine **matdat1** compute and record the results of the following matrix expressions in the space provided. If an operation is not defined, state why.

$$A = \begin{bmatrix} 5 & -2 & 1 \\ 1 & 0 & 4 \\ -3 & 7 & 2 \end{bmatrix} \quad B = \begin{bmatrix} 2 & 2 & 3 \\ -1 & 4 & 1 \\ 5 & -3 & 0 \end{bmatrix} \quad C = \begin{bmatrix} 1 & -1 & 2 \\ 0 & 1 & 4 \\ -5 & 3 & 6 \end{bmatrix}$$

$$D = \begin{bmatrix} -1 & 2 & 3 \\ 0 & 4 & 5 \end{bmatrix} \quad x = \begin{bmatrix} -2 \\ 3 \\ 1 \end{bmatrix}$$

$A + B =$ $\qquad\qquad$ $B - D =$

$A * B =$ $\qquad\qquad$ $B * A =$

$D * C =$ $\qquad\qquad$ $C' =$

$C * x =$ $x * x =$

$x' * x =$ $((A - B) * x)' =$

$A^2 =$ $A * A =$

$6 * D =$ $5 * A - 3 * B =$

Enter each of the following matrices into MATLAB . Exercises 2-3 refer to these matrices.

$$A = \begin{bmatrix} 1 & 3 \\ 2 & 4 \\ 3 & 1 \end{bmatrix} \quad B = \begin{bmatrix} -1 & 2 \\ 4 & -2 \\ 7 & -1 \end{bmatrix} \quad C = \begin{bmatrix} 1 & 5 \\ -5 & 3 \end{bmatrix} \quad D = \begin{bmatrix} 4 & 3 & -2 \\ 1 & 0 & 5 \\ 2 & -1 & 6 \end{bmatrix}$$

2. Perform the following matrix algebra computations in MATLAB . Record your result below the expression.

a) $A + B$ b) $B + C$ c) $D * A$ d) $2 * A - 3 * B$ e) A' f) C^2

3. Perform each of the following matrix algebra statements in MATLAB . Briefly describe the action taken in each part. Warning: these are not the standard matrix operations.

a) $A. * B$ ________________________

b) $A./B$ ________________________

c) $A.\hat{}3$ ________________________

4. Enter each of the following matrices into MATLAB . To enter a complex number like $2 - 3i$ type $2 - 3 * i$ (or $2 - 3i$) with no spaces intervening. Then compute the indicated matrix algebra statements in MATLAB . Record your result below the expression.

$$E = \begin{bmatrix} 1+i & 3 \\ 2-i & 4+2i \\ 3-i & 1+i \end{bmatrix} \quad F = \begin{bmatrix} -1 & 2-3i \\ 4+i & -2 \\ 7-i & -1 \end{bmatrix} \quad G = \begin{bmatrix} 1+i & 5+2i \\ 2-5i & 3-4i \end{bmatrix}$$

a) $E+F$ b) $F+G$ c) $G+E$

d) $2*E-3*F$ e) E' f) G^2

5. Let A and X be the matrices defined below.

$$A = \begin{bmatrix} 6 & -1 & 1 \\ 0 & 13 & -16 \\ 0 & 8 & -11 \end{bmatrix} \qquad X = \begin{bmatrix} 10.5 \\ 21.0 \\ 10.5 \end{bmatrix}$$

a) Determine a scalar r such that $AX = rX$. $r =$ ________________

b) Compute $AX - rX$ for the value of r from part a). ________________

c) Is it true that $A'X = rX$ for the value of r determined in part a)?

(Circle one: Yes No)

6. Let A and X be the matrices defined below.

$$A = \begin{bmatrix} -7.5 & 8.0 & 16.0 \\ -2.0 & 2.5 & 4.0 \\ -2.0 & 2.0 & 4.5 \end{bmatrix} \qquad X = \begin{bmatrix} 4 \\ 1 \\ 1 \end{bmatrix}$$

a) Determine a scalar r such that $AX = rX$. $r =$ ________________

b) Compute $AX - rX$ for the value of r from part a). ________________

c) Is it true that $A'X = rX$ for the value of r determined in part a)?

(Circle one: Yes No)

7. Use MATLAB to evaluate each polynomial $p(x)$ at the endpoints of the given interval, draw the graph of $p(x)$ over the interval using the procedure in Example 1, and evaluate $p(x)$ at the given matrix.

a) $p(x) = 2x^2 - x + 1$, $[-2, 2]$, $A = \begin{bmatrix} 1 & 0 & 0 & 0 \\ 2 & 3 & 0 & 0 \\ 4 & 5 & 6 & 0 \\ 7 & 8 & 9 & 0 \end{bmatrix}$

$p(-2) =$ ____________

$p(2) =$ ____________

$p(A) =$ ____________

b) $p(x) = x^3 - 2x^2 + 2$, $[-1, 3]$, $A = \begin{bmatrix} 1 & 2 \\ 3 & 4 \end{bmatrix}$

$p(-1) =$ ____________

$p(3) =$ ____________

$p(A) =$ ____________

8. Every square matrix A has a polynomial associated with it called the **characteristic polynomial**[1] of A, which can be obtained using the MATLAB command **poly(A)**. (Warning: do not use command **polyval**.)

[1] In Lab 13 we formally define the characteristic polynomial of a matrix and use it to determine particular information about a matrix

a) Let $A=\begin{bmatrix} 5 & -8 & -1 \\ 4 & -7 & -4 \\ 0 & 0 & 4 \end{bmatrix}$.

What is the characteristic polynomial $p(x)$ of A? ____________________

Evaluate $p(A)$ ____________________.

b) Let $A=\begin{bmatrix} 7 & -4 & 0 \\ 8 & -5 & 0 \\ -4 & 4 & 3 \end{bmatrix}$.

What is the characteristic polynomial $p(x)$ of A? ____________________

Evaluate $p(A)$ ____________________.

c) Use the following commands to randomly generate a 3×3 matrix A with entries which are whole numbers.

```
rand('state', sum(100 * clock))
A=fix(10*rand(3))
```

What is the characteristic polynomial $p(x)$ of A? ____________________

Evaluate $p(A)$ ____________________.

d) Make a conjecture about the result $p(A)$, where A is any square matrix and $p(x)$ is its characteristic polynomial.

____

e) Verify your conjecture in part (d) on a 4×4 matrix B and on a 5×5 matrix C. Use commands **B=fix(10*rand(4))** and **C=fix(10*rand(5))**, respectively, to obtain randomly generated matrices. Was your conjecture verified? Explain.

____

9. Evaluate the polynomial $p(x) = 5x^3 + 3x^2 - 2x + 4$ for $A=\begin{bmatrix} 1 & 2 & 3 \\ 0 & 0 & a \\ 0 & 0 & 0 \end{bmatrix}$ by hand:

$p(A) =$ ______________

Verify your answer by using appropriate MATLAB commands involving symbolic matrices as in Lab 2.3. Record the MATLAB commands you use below.

____

Section 3.2

Generating Matrices

The $n \times n$ identity matrix is denoted I_n. MATLAB has a command to generate I_n when it is needed. The command **eye** behaves as follows:

eye(2)	displays the **2** × **2** identity matrix
eye(5)	displays the **5** × **5** identity matrix
t=10;eye(t)	displays the **10** × **10** identity matrix
eye(size(A))	displays the identity matrix the same size as A

Two other MATLAB commands, **zeros** and **ones**, behave in a similar manner. The command **zeros** produces a matrix of all zeros, while command **ones** generates a matrix of all ones. Rectangular matrices of size $m \times n$ can be generated using commands

eye(m,n), zeros(m,n), ones(m,n)

where m and n have been previously defined with positive integer values in MATLAB . For example to generate a column with four zeros use command

zeros(4,1)

or

m=4;n=1;zeros(m,n)

MATLAB can generate random numbers in several ways. Type **rand** and then use your up-arrow key to repeat the command several times. Command **rand** produces random values in the interval (0,1). Command **randn;** uses a random number generator that produces values on either side of zero in a manner known as normally distributed with variance one. (See **help rand** and **help randn** for more details.) Type **randn** and repeat the **randn** command until you get a value greater than 2 or less than -2.

The commands **rand** and **randn** have variations just like those for **eye, ones**, and **zeros**. To experiment type the following commands and then construct several of your own.

rand(5)
rand(4, 1)
rand(3, 6)
rand(size(eye(3)))

In our work it is often convenient to be able to easily generate matrices for use in exercises or to check conjectures about matrix properties. The **rand** and **randn** commands give us real matrices usually with entries which are not whole numbers. The command

fix(rand(5))

generates a $\mathbf{5} \times \mathbf{5}$ matrix with integer entries which are obtained by rounding the entries of the matrix produced by **rand(5)** to the nearest integer towards zero. (For more information on **fix** use **help**.) Often the matrix produced by the command **fix(rand(k))**, where k is an integer denoting the size of the matrix desired, contains many zeros. One way to obtain fewer zeros is to multiply each element by **10** before 'fixing' it. Command

fix(10∗rand(5))

performs this task. To obtain a $\mathbf{5} \times \mathbf{5}$ complex matrix type the commands

i = sqrt(−1); C = fix(10∗rand(5)) + i∗fix(10∗rand(5))

Use the arrow keys to recall the previous command and edit it to obtain a $\mathbf{3} \times \mathbf{3}$ complex matrix. The command **randn** can be used in place of **rand** in the preceding discussion.

Exercises 3.2

1. Enter each of the following matrices into MATLAB .

$$C = \begin{bmatrix} 1 & 5 \\ -5 & 3 \end{bmatrix} \qquad D = \begin{bmatrix} 4 & 3 & -2 \\ 1 & 0 & 5 \\ 2 & -1 & 6 \end{bmatrix}$$

Enter each of the following MATLAB commands and carefully analyze the display generated. Make certain you understand the behavior of each command. For new commands use **help** for a brief description. Write a brief description of the action of the commands in the space provided.

a) **5*eye(2)** b) **eye(2) + ones(2)**

c) **ones(size(C)), zeros(size(C)), C + ones(size(C))**

d) **D, diag(D), diag(diag(D))** e) **diag([-3 4]), diag([5 -7 1])**

f) **D, triu(D)** g) **D, tril(D)**

a) ______________________

b) ______________________

c) ______________________

d) ______________________

e) ______________________

f) ______________________

g) ______________________

2. In each of the following construct a MATLAB command to generate the matrix described. For example, a row with five ones is generated by **ones(1,5)**. Record your command in the space provided. (Do not explicitly type the entries of the matrix described.)

a) A column with 8 ones. ______________________

b) A row with 10 threes. ______________________

c) A $\mathbf{5 \times 5}$ matrix with a diagonal of all sevens. ______________________

d) The matrix $\begin{bmatrix} \mathbf{2} & \mathbf{1} & \mathbf{1} \\ \mathbf{1} & \mathbf{2} & \mathbf{1} \\ \mathbf{1} & \mathbf{1} & \mathbf{2} \end{bmatrix}$ ______________________

e) The matrix $\begin{bmatrix} \mathbf{5} & \mathbf{-1} & \mathbf{-1} \\ \mathbf{-1} & \mathbf{5} & \mathbf{-1} \\ \mathbf{-1} & \mathbf{-1} & \mathbf{5} \end{bmatrix}$ ______________________

3. Construct a MATLAB command to generate the $\boldsymbol{n \times n}$ matrix $\boldsymbol{A}$ which has $\boldsymbol{a_{ij} = 1}$ for $\boldsymbol{i \neq j}$, $\boldsymbol{a_{ii} = 1 - n}$. Demonstrate your command for cases $\boldsymbol{n = 3, 5, 8}$. Record your command below.

____

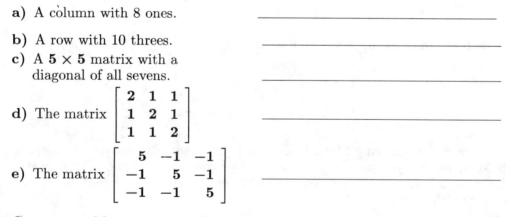

4. Repeat Exercise 3, but for matrix $\boldsymbol{B}$ which has $\boldsymbol{b_{ij} = 1}$ for $\boldsymbol{i \neq j}$, $\boldsymbol{b_{ii} = 1/n}$.

---

5. The description for generating random matrices with integer entries given above used the **fix** command. MATLAB has commands **ceil**, **floor**, and **round** which can be used in place of **fix**. Use **help** to obtain a description of these commands.

a) Apply each of the commands to values **2.6**, **3.2**, and **−1.5** to observe their behavior.

b) Generate several random integer matrices using each of these commands.

6. Generate matrix $\boldsymbol{A}$ from command **A** = **ceil(10∗rand(5))**. Use MATLAB to verify each of the following.

a) $\boldsymbol{S = A + A'}$ is symmetric.
b) $\boldsymbol{T = A - A'}$ is skew symmetric.
c) $\boldsymbol{A}$ is the same as $\frac{1}{2}\boldsymbol{S} + \frac{1}{2}\boldsymbol{T}$.
d) For **L** = **tril(A,−1)**, **D** = **diag(diag(A))**, and **U** = **triu(A,1)**, **L** + **D** + **U** = **A**.

7. Experiment: Investigate the form of the product of two diagonal matrices.

a) In MATLAB do the following: **A** = **diag** ([**1 2 7**]), **B** = **diag**([**−3 4 2**]).
Compute **A∗B**. Inspect the 'form' of the result.
Complete the following statement: **A∗B** is a ______________ matrix.

b) In MATLAB do the following:
A = **diag**([**4 3 2 −1**]), **B** = **diag**([**0 1 5 −3**]). Compute **A∗B**.
Inspect the 'form' of the result.
Complete the following statement: **A∗B** is a ______________ matrix.

c) Conjecture: The product of two diagonal matrices is a ______________ matrix.
(Try to prove your conjecture for n×n diagonal matrices.)

8. Experiment: Investigate the form of the product of two lower triangular matrices.

a) In MATLAB do the following:
A = **tril(fix(10∗rand(3)))**, **B** = **tril(fix(10∗rand(3)))**. Compute **A∗B**.
Inspect the 'form' of the result.
Complete the following statement: **A∗B** is a ______________ matrix.

b) In MATLAB do the following:
A = tril(fix(10*rand(5))), **B = tril(fix(10*rand(5)))**. Compute **A*B**.
Inspect the 'form' of the result.
Complete the following statement: **A*B** is a ______________ matrix.

c) Conjecture: The product of two lower triangular matrices is a ______________ matrix. (Try to prove your conjecture for n×n lower triangular matrices.)

9. In Exercise 8 replace lower triangular by upper triangular and **tril** by **triu** and repeat the experiments. Conjecture: The product of two upper triangular matrices is a ______________ matrix.

10. Experiment: Investigate the form of the product of m×n matrix $\boldsymbol{A}$ times column $\boldsymbol{x}$ of size $\boldsymbol{n} \times \mathbf{1}$.

a) In MATLAB let $\boldsymbol{A} = \begin{bmatrix} 2 & -4 \\ 0 & 1 \\ -3 & 0 \end{bmatrix}$ and $\boldsymbol{x} = \begin{bmatrix} 5 \\ -8 \end{bmatrix}$. Compute $\boldsymbol{A} * \boldsymbol{x}$.

$$\boldsymbol{A} * \boldsymbol{x} = k_1 \begin{bmatrix} 2 \\ 0 \\ -3 \end{bmatrix} + k_2 \begin{bmatrix} -4 \\ 1 \\ 0 \end{bmatrix} = k_1 \mathrm{col}_1(\boldsymbol{A}) + k_2 \mathrm{col}_2(\boldsymbol{A})$$

Determine k_1 and k_2. $k_1 =$ ______________ $k_2 =$ ______________

b) In MATLAB let $\boldsymbol{A} = \begin{bmatrix} 1 & 0 & 4 \\ 2 & 1 & -1 \\ 3 & 2 & 5 \end{bmatrix}$ and $\boldsymbol{x} = \begin{bmatrix} 0 \\ 3 \\ -2 \end{bmatrix}$. Compute $\boldsymbol{A} * \boldsymbol{x}$.

$$\boldsymbol{A} * \boldsymbol{x} = k_1 \begin{bmatrix} 1 \\ 2 \\ 3 \end{bmatrix} + k_2 \begin{bmatrix} 0 \\ 1 \\ 2 \end{bmatrix} + k_3 \begin{bmatrix} 4 \\ -1 \\ 5 \end{bmatrix} = k_1 \mathrm{col}_1(\boldsymbol{A}) + k_2 \mathrm{col}_2(\boldsymbol{A}) + k_3 \mathrm{col}_3(\boldsymbol{A})$$

Determine k_1, k_2 and k_3. $k_1 =$ ________ $k_2 =$ ________ $k_3 =$ ________

c) Matrix $\boldsymbol{A} = (a_{ij})$ is 3×4 and column $\boldsymbol{x} = (x_k)$ is 4×1. Express the product $\boldsymbol{A} * \boldsymbol{x}$ in terms of the elements of $\boldsymbol{x}$ and the columns of $\boldsymbol{A}$.

$\boldsymbol{A} * \boldsymbol{x} =$ ____

d) Conjecture: The product $\boldsymbol{A} * \boldsymbol{x}$ is a sum of the ______________ of $\boldsymbol{A}$ with coefficients which are the ______________ of $\boldsymbol{x}$. (Try to prove this in general.)

Section 3.3

Display Formats

MATLAB stores matrices in a decimal form and does its arithmetic computations using a decimal-type arithmetic. This decimal form retains about 16 digits, but not all digits must be shown. In between what goes on in the machine and what is shown on the screen are routines that convert or format the numbers into displays. Here we give an overview of the display formats that we will use. (For more information type **help format**.)

If the matrix contains all integers then the entire matrix is displayed as integer values; that is, no decimal points appear.

If any entry in the matrix is not exactly represented as an integer, then the entire matrix is displayed in what is known as **format short. Such a display shows 4 places behind the decimal point** and the last place may have been rounded. The exception to this is zero. **If an entry is exactly zero, then it is displayed as an integer zero.** Enter the matrix

$$Q = [\ \mathbf{5}\quad \mathbf{0}\quad \mathbf{1/3}\quad \mathbf{2/3}\quad \mathbf{7.123456}\quad \mathbf{.00000197}\ ]$$

into MATLAB . The display is

```
Q =

    5.0000         0    0.3333    0.6667    7.1235    0.0000
```

To see more than 4 places, we change the display format. One way to proceed is to use command

format long

which shows 15 places The matrix Q above in format long is

```
Q =

  Columns 1 through 4

5.00000000000000                  0   0.33333333333333   0.66666666666667

  Columns 5 through 6

7.12345600000000   0.00000197000000
```

There are other display formats that use an exponent of 10. They are **format short e** and **format long e**. The 'e-formats' are often used in numerical analysis. Try these formats and display matrix $\boldsymbol{Q}$.

MATLAB can display values in a fraction form using format **rat**, short for rational display. To make rational form easier to display we have written another command, **rational**, that produces the same result as format **rat** shown above.

Inspect the output from the following sequence of MATLAB commands.

format short
V= [1 1/2 1/6 1/12]

displays

```
V =

    1.0000    0.5000    0.1667    0.0833
```

The command

rational(V)

displays

```
ans =

    1    1/2    1/6    1/12
```

> Warning: Rational output is displayed in what is called string form. **Strings cannot be used with arithmetic operators.** Thus rational output is for 'looks' only.

When MATLAB starts, the format in effect is **format short**. If you change the format, it remains in effect until another format command is executed. Some MATLAB routines change the format within the routine. For further information use **help format**.

> Since MATLAB has pull-down menus, changing formats can be done choosing the Options menu and then Numeric format.

Exercises 3.3

1. Enter matrix $\boldsymbol{A} = \begin{bmatrix} \mathbf{2/3} & \mathbf{5/8} & \mathbf{11/3} \\ \mathbf{4/5} & \mathbf{3} & \mathbf{-5} \end{bmatrix}$ into MATLAB . Execute the following sequence of commands to observe the various format displays. Carefully note the differences in these display formats.

a) **format short, A** b) **format short e, A** c) **format long, A**

d) **format long e, A** e) **rational(A)**

2. Enter the following commands and observe how the display of the columns vary with the long and short formats.

format short, A = rand(8)
format short e, A
format long, A
format long e, A
format short

3. Use command **help hilb**. Display the $\mathbf{5 \times 5}$ Hilbert matrix in each of the display formats. Note that the format short display rounds to get the displayed fourth decimal place.

Section 3.4

An Application: Population Growth

Suppose a species of animal can live to a maximum age of two years. Suppose also that the number of males is a fixed percentage of the female population. Thus, in studying population growth of this species we can ignore the male population and concentrate on the female population. We shall keep track of the number of living females of ages $\mathbf{0, 1, 2}$ (age zero means age less than 1) by defining a **population vector** containing entries that represent the population of each age group in millions.

Let the initial population vector be

$$IP = \begin{bmatrix} \mathbf{14.5} \\ \mathbf{15.3} \\ \mathbf{11.3} \end{bmatrix}$$

This means that initially there are 14.5 million females of age 0, 15.3 million of age 1, and 11.3 million of age 2.

In order to determine how the population changes we need to know the birth and survival rates for each age class of females. This information for our animal is as follows.

Births:

\# of new age zero =
.559∗# of current age zero **+.6∗** # of current age one **+.1∗** # of current age two

Survivors:

$$\# \text{ of new age one} = \mathbf{.7}* \# \text{ of current age zero}$$

$$\# \text{ of new age two} = \mathbf{.3}* \# \text{ of current age one}$$

The numbers .559, .6, and .1 are the birth rates for the age groups respectively and the numbers .7 and .3 are the survival rates for ages zero and one respectively. This information is modeled by the system of three equations given above. In matrix form this system is

$$\begin{bmatrix} \mathbf{.559} & \mathbf{.6} & \mathbf{.1} \\ \mathbf{.7} & \mathbf{0} & \mathbf{0} \\ \mathbf{0} & \mathbf{.3} & \mathbf{0} \end{bmatrix} \begin{bmatrix} \# \text{ of current age zero} \\ \# \text{ of current age one} \\ \# \text{ of current age two} \end{bmatrix} = \begin{bmatrix} \# \text{ of new age zero} \\ \# \text{ of new age one} \\ \# \text{ of new age two} \end{bmatrix}$$

Call the coefficient matrix above **A**. The matrix **A** is called a **transition matrix**. It follows that the population vector after one year is given by

$$\mathbf{PT1} = \mathbf{A} * \mathbf{IP} \tag{3.1}$$

Perform this operation in MATLAB and record the result below.

$$\mathbf{PT1} = \begin{bmatrix} \quad \\ \quad \\ \quad \end{bmatrix}$$

Similarly the population vector after two years is

$$\mathbf{PT2} = \mathbf{A} * \mathbf{PT1} \tag{3.2}$$

Perform this operation in MATLAB and record the result below.

$$\mathbf{PT2} = \begin{bmatrix} \quad \\ \quad \\ \quad \end{bmatrix}$$

An alternate way to get vector **PT2** is to substitute (3.1) into (3.2):

$$\mathbf{PT2} = \mathbf{A}*\mathbf{PT1} = \mathbf{A} * (\mathbf{A} * \mathbf{IP}) = \mathbf{A}^2\mathbf{IP}$$

Thus

$$\mathbf{PT2} = \mathbf{A}^2\mathbf{IP} \tag{3.3}$$

The expression in (3.3) gives PT2 in terms of the transition matrix **A** and the initial population vector **IP**. Continuing in this way, substitute (3.3) into the formula **PT3 = A*PT2** to get

$$\mathbf{PT3} = \mathbf{A}^3\mathbf{IP} \tag{3.4}$$

By following this pattern of substitution the population vector after N years, denoted **PTN** is:

$$\mathbf{PTN} = \mathbf{A}^N\mathbf{IP} \tag{3.5}$$

To use equation (3.5) in MATLAB with N = 3 type

PT3 = A^3*IP

Perform this operation and record the result.

$$\mathbf{PT3} = \begin{bmatrix} \\ \\ \\ \end{bmatrix}$$

Next use the up arrow ↑ to recall the command and then edit it to calculate population vectors at other times. Replace the exponent 3 by 4. Then

$$\mathbf{PT4} = \begin{bmatrix} \\ \\ \\ \end{bmatrix}$$

Repeat the procedure to compute

$$\mathbf{PT5} = \begin{bmatrix} \\ \\ \\ \end{bmatrix}$$

In MATLAB form the population matrix

PM = [PT1 PT2 PT3 PT4 PT5]

Do the entries in the rows of **PM** seem to be forming a pattern? (Circle your choice.)

YES NO UNSURE

Look at the first row of **PM**. The entries represent the number of females of age zero for five successive years. Do these numbers seem to be stabilizing?

YES NO UNSURE

To gain a geometric perspective on the last two questions type

plot(PM′)

This particular snapshot indicates the following tendencies on the right-hand side of the screen. The top curve, which represents the population of females of age zero, is nearly horizontal, indicating that it is approaching a stable population. But the middle curve, which represents females of age one, seems to be decreasing slightly, while the bottom curve, which represents females of age 2, seems to be increasing slightly.

To gain a broader view, let's see what happens over the next 5 years; that is a total of 10 years. Press any key to get back to the command screen. Use the up arrow ↑ to edit command lines to compute **PT6, PT7, PT8, PT9,**and **PT10**. Next append these additional population vectors to **PM** by typing command

PM = [PM PT6 PT7 PT8 PT9 PT10]

Type the command **plot(PM′)** again. Notice that now each of the curves levels out towards the right, indicating that each population is reaching a 'steady state'.

What are these steady states?

To answer this question press a key to return to the command screen. The population matrix **PM** should still be displayed. (If not type **PM**.) Examine the first row. The entries form a sequence. What appears to be the limit of this sequence?

____

What appears to be the limit of the sequence of second entries?

____

What appears to be the limit of the sequence of third entries?

____

To test your conjecture compute **PT15** and **PT20**. From this evidence, what conjecture would you make for the steady state population vector?

$$\begin{bmatrix} \\ \\ \\ \end{bmatrix}$$

Recall that MATLAB has display modes for the screen information. Change the mode to **format long** and display **PT15** and **PT20**. Would you change your conjecture?

What then is the steady state population? We will show how to determine this later. However, experiment to determine it using the tools you have now. State your result below.

Exercises 3.4

1. Let $\boldsymbol{A}$ be the transition matrix and $\boldsymbol{IP}$ be the initial population vector defined below.

$$A = \begin{bmatrix} .21 & .64 & .12 \\ .69 & 0 & 0 \\ 0 & .36 & 0 \end{bmatrix} \qquad IP = \begin{bmatrix} 14.5 \\ 15.3 \\ 11.3 \end{bmatrix}$$

Compute **PT10**: $\begin{bmatrix} \quad \end{bmatrix}$ Compute **PT11**: $\begin{bmatrix} \quad \end{bmatrix}$

Does it appear that the process is reaching a steady state? Circle one:

YES NO UNSURE

2. Let $\boldsymbol{A}$ be the transition matrix and $\boldsymbol{IP}$ be the initial population vector defined below.

$$A = \begin{bmatrix} .868 & .4 & .2 \\ .3 & 0 & 0 \\ 0 & .2 & 0 \end{bmatrix} \qquad IP = \begin{bmatrix} 14.5 \\ 15.3 \\ 11.3 \end{bmatrix}$$

Compute **PT10**: $\begin{bmatrix} \quad \end{bmatrix}$ Compute **PT11**: $\begin{bmatrix} \quad \end{bmatrix}$

Does it appear that the process is reaching a steady state? Circle one:

YES NO UNSURE

3. A retail store R has 55% of the market versus 45% for the lone competitor C. An advertisement firm predicts that an ad campaign will draw 20% of C's customers to R, while R will lose only 15% of their own customers to C. Write the initial population vector $\boldsymbol{IP}$.

$$\boldsymbol{IP} = \begin{bmatrix} \\ \\ \end{bmatrix}$$

Fill in the entries of the transition matrix $\boldsymbol{A}$, where $\boldsymbol{A(i,j)}$ is the percentage share that is drawn from store i to store j.

$$\boldsymbol{A} = \begin{bmatrix} & \\ & \end{bmatrix}$$

Compute **PT10**: $\begin{bmatrix} \\ \\ \end{bmatrix}$ Compute **PT11**: $\begin{bmatrix} \\ \\ \end{bmatrix}$

Does it appear that the process is reaching a steady state? (Circle one.)

YES NO UNSURE

4. Every day a manager rates the performance of each member of her staff as poor, average, or excellent. If a worker was rated poor on one day then the probability that on the next day the worker will be rated poor is .2, average is .7, and excellent is .1. If a worker was rated average on one day then probability that on the next day the worker will be rated poor is .3, average is .4, and excellent is .3. If a worker was rated excellent on one day then the probability that on the next day the worker will be rated poor is .1, average is .7, and excellent is .2. Initially 25% of the workers were excellent, 65% average, and 10% poor.

(i) What percentage was rated excellent after 30 days?

(ii) What percentage was rated excellent after 365 days?

(iii) What percentage will be rated excellent in the long run?

(Hint: Complete the construction of the transition matrix below, then form the initial distribution. Proceed as in the population model given in this section.)

		Today P	Today A	Today E
Tomorrow	P	.2		
	A	.7		
	E	.1		

Transition Matrix

P = poor
A = average
E = excellent

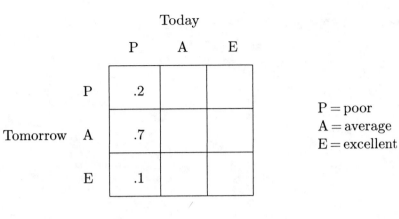

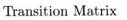

<< NOTES; COMMENTS; IDEAS >>

LAB **4**

Homogeneous Systems, Echelon Forms, and Inverses

Topics: solving homogeneous systems using **rowop**; finding general solutions; seeing the reduced row echelon form as a movie and then step-by-step; using **rowop** to get the reduced row echelon form; inverses from reduced row echelon form computations; inverses; command **invert**.

Introduction

This lab is crucial because it introduces routines that will be used in every other lab and project. Moreover, it explains and illustrates the algorithmic nature of the reduction process for converting a matrix to reduced row echelon form.

Section 4.1 uses the routine **rowop**, introduced in Lab 2, to write the general solution of a homogeneous system of linear equations. This routine is the first step in the automation of the reduction process that was first carried out with pencil and paper. When using **rowop** the student can concentrate on the strategy and leave the arithmetic to MATLAB .

Section 4.2 presents the middle and final steps. The middle step involves the routine **rrefquik**, which allows the student to view the algorithmic nature of the process of reduction without being confronted with any strategic decisions. If **rrefquik** is regarded as a video on reduction, then **rrefview** and **rrefstep** present the viewer with two alternate, slow-motion performances. In fact **rrefview**, which shows reduction frame-by-frame, provides a type of computer visualization in a very unusual context.

The final step is MATLAB routine **rref**, which immediately computes and displays the reduced row echelon form. (**rref** is a *black-box*, that is, a routine which presents an answer with no hint of the underlying process.) We could have begun with this routine but our aim is for the student to understand the workings of the reduction process before taking this giant step. Section 4.3 shows how to apply **rref** to compute the inverse of a matrix. Here too we could have taken a black-box approach by introducing an appropriate MATLAB command but we chose instead to present a method that reinforces the meaning of the inverse of a matrix.

Ultimately the routine **rref** will become a student's most useful tool. It is used to study every concept in the rest of the course. However, we will see situations (for instance, in Lab 8 on determinants) where **rowop** is more appropriate.

Section 4.1

Homogeneous Systems

To solve the homogeneous linear system

$$\begin{aligned} 2x_1 - 3x_2 - 5x_3 &= 0 \\ x_1 - 2x_2 - 4x_3 &= 0 \\ -3x_1 + 4x_2 + 6x_3 &= 0 \end{aligned}$$

in MATLAB execute routine **rowop** and enter the augmented matrix

$$\begin{bmatrix} 2 & -3 & -5 & 0; & 1 & -2 & -4 & 0; & -3 & 4 & 6 & 0 \end{bmatrix}$$

Select appropriate row operations until you produce the following equivalent augmented matrix:

$$\begin{bmatrix} 1 & 0 & 2 & | & 0 \\ 0 & 1 & 3 & | & 0 \\ 0 & 0 & 0 & | & 0 \end{bmatrix}$$

To write out the **general solution** of the corresponding homogeneous system proceed as follows. The 'reduced' homogeneous system is

$$\begin{aligned} x_1 \qquad + 2x_3 &= 0 \\ x_2 + 3x_3 &= 0 \end{aligned}$$

There are 3 unknowns, but only two equations. Hence one unknown can be chosen arbitrarily. Usually the unknown(s) in columns without leading ones are chosen arbitrarily. Here, let

$$x_3 = r$$

where r is any real number. Then working from the last equation upward we have

$$\begin{aligned} x_2 &= -3x_3 = -3r \\ x_1 &= -2x_3 = -2r \end{aligned}$$

Hence the general solution (that is, the set of all solutions) is given by

$$\mathbf{X} = \begin{bmatrix} x_1 \\ x_2 \\ x_3 \end{bmatrix} = \begin{bmatrix} -2r \\ -3r \\ r \end{bmatrix} = r \begin{bmatrix} -2 \\ -3 \\ 1 \end{bmatrix}$$

For practice find the general solution of

$$\begin{aligned} 3x_1 + x_2 + 2x_3 &= 0 \\ -4x_1 \qquad + x_3 &= 0 \\ 2x_1 + 2x_2 + 5x_3 &= 0 \end{aligned} \qquad \text{and} \qquad \begin{aligned} -2x_1 + 3x_2 + 4x_3 + 4x_4 &= 0 \\ x_1 \qquad + 2x_3 \qquad &= 0 \\ 5x_1 - 6x_2 - 6x_3 - 8x_4 &= 0 \end{aligned}$$

Record your solutions here:

General solution of first system:

General solution of second system:

Exercises 4.1

1. Use **rowop** to find the general solution of the following homogeneous system of linear equations. Record your solution next to the linear system.

$$\begin{aligned} 3x_1 - 2x_2 + 12x_3 &= 0 \\ x_1 + x_2 - x_3 &= 0 \\ 2x_1 + x_2 + x_3 &= 0 \end{aligned}$$

2. Usc **rowop** to find the general solution of the following homogeneous system of linear equations. Record your solution next to the linear system.

$$\begin{aligned} -2x_1 + 3x_2 + 4x_3 + 4x_4 &= 0 \\ x_1 - 2x_3 &= 0 \\ 5x_1 - 2x_2 + 4x_3 - 8x_4 &= 0 \end{aligned}$$

3. Use **rowop** to find the general solution of the following homogeneous system of linear equations. Record your solution next to the linear system.

$$\begin{aligned} x_1 + 2x_2 + 3x_3 &= 0 \\ 4x_1 + 5x_2 + 6x_3 &= 0 \\ 7x_1 + 8x_2 &= 0 \end{aligned}$$

4. Use **rowop** to find the general solution of the following homogeneous system of linear equations. Record your solution next to the linear system.

$$\begin{aligned} x_1 - x_2 + 2x_3 + x_5 &= 0 \\ 2x_1 + x_2 + x_3 + x_4 + x_5 &= 0 \\ x_1 + x_2 + 2x_4 + 2x_5 &= 0 \end{aligned}$$

The routine **symrowop**, introduced in Section 2.3, can be used to solve homogeneous linear systems in which coefficients contain a parameter. In Exercises 5 to 7 use **symrowop** and record the restrictions you acknowledge. From these restrictions determine which values of parameter c produce a system with more than one solution. Hint: substitute each value of c into the system to form the corresponding numeric matrix and use **rowop** to solve the linear system.

5. $$\begin{aligned}(c+2)x_1 + 2x_2 &= 0\\ 2x_1 + x_2 &= 0\end{aligned}$$

Restrictions: ______________________

$c =$ ______________________

6. $$\begin{aligned}3x_1 + \qquad 12x_2 &= 0\\ -2x_1 + (c^2-9)x_2 &= 0\end{aligned}$$

Restrictions: ______________________

$c =$ ______________________

7. $$\begin{aligned}cx_1 + x_2 + \qquad 3x_3 &= 0\\ 2x_1 + x_2 + \qquad 3x_3 &= 0\\ 4x_1 + 2x_2 + (c+9)x_3 &= 0\end{aligned}$$

Restrictions: ______________________

$c =$ ______________________

Section 4.2

Reduced Row Echelon Form

The reduced row echelon form of a matrix is obtained by the systematic application of row operations to transform the matrix to be as close to an identity matrix as possible. Hence row operations are chosen to introduce as many zeros as possible. As the reduction (or elimination) process progresses zeros are obtained above and below certain matrix entries. Such entries are called **pivots** or **pivot elements** and the row containing them is called the **pivot row**.

To illustrate the 'zeroing out' of entries in the process of obtaining the reduced row echelon form of a matrix execute the following MATLAB commands. This demonstration should give you a feel for the process.

```
A = fix(10*rand(10,6))
rrefquik(A)
```

Repeat this demonstration several times. (To see the reduction with rational displays, precede the **rrefquik** command with **format rat**; afterwards type **format** to reset the format to short.) Note that many columns eventually look like columns of an identity matrix. MATLAB is automatically choosing row operations to eliminate entries and obtain an equivalent matrix which is in reduced row echelon form. (The phrase "reduced row echelon form" is often abbreviated rref.) That is, the final matrix satisfies the following properties.

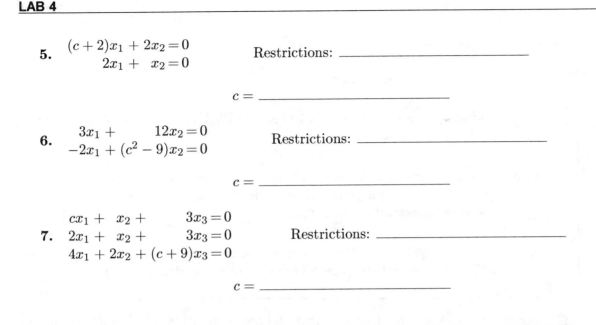

- All zero rows, if any, come last.
- The first entry of a nonzero row is 1.(This is called a **leading 1** of the row.)
- In each nonzero row, the leading 1 appears to the right and below the leading 1 in preceding rows. (Eventually the leading 1's appear as a staircase downward to the right.)
- Any column in which a leading 1 appears has zeros in every other entry.(That is, such columns are columns of some identity matrix.)

To illustrate more slowly the reduction to rref and the properties listed above enter the following MATLAB commands. The routine **rrefview** gives a frame-by-frame look at the reduction process.

A = [0 0 0 0;1 2 3 4;1 2 −5 6;3 0 2 1]
rrefview(A)

In order to produce the rref without MATLAB routines making the decisions we use **rowop**. Routine **rowop** requires that you choose row operations to produce the reduced row echelon form or the row echelon form. **rowop** performs the arithmetic of the row operations for you so that you may concentrate on the logical decisions involved. Use **rowop** to find the rref of the following matrix $\boldsymbol{A}$ and record your result next to $\boldsymbol{A}$.

$$\boldsymbol{A} = \begin{bmatrix} 3 & 0 & 2 & 1 \\ 1 & 2 & 3 & 4 \\ 1 & 2 & -5 & 6 \end{bmatrix} \qquad \begin{bmatrix} \quad & \quad & \quad & \quad \\ \\ \\ \end{bmatrix}$$

To check your results you can use **rrefquik** or **rrefview**.

Warning: When using **rowop** it is often convenient to use expressions for the multipliers. Within **rowop** the matrix being transformed is named $\boldsymbol{A}$, regardless of the name assigned to the matrix outside of **rowop**. Hence a multiplier could be written in the form $-\boldsymbol{A}(2,1)/\boldsymbol{A}(1,1)$. This is especially valuable when the entries are in decimal form. This procedure will assure that you are using the correct value and not a four-decimal approximation.

Find the rref of matrix **C** below using **rowop** and construct the multipliers for the row operations using the convention discussed above. Record the result in the space provided.

$$\mathbf{C} = \begin{bmatrix} 3 & -1 & 2 \\ 1 & 2 & 4 \\ -1 & 5 & 6 \\ 4 & 1 & 6 \end{bmatrix} \qquad \text{rref} \quad \text{of} \quad \mathbf{C} = \begin{bmatrix} \quad & \quad & \quad \\ & & \\ & & \\ & & \end{bmatrix}$$

There are many different ways to choose row operations to apply to matrix $\boldsymbol{A}$ to obtain its reduced row echelon form, but the end result must be the same. The reduced row echelon form of a matrix $\boldsymbol{A}$ is unique. If you need help in practicing the selection of row operations to get the reduced row echelon form of a matrix, use **help** to obtain a description of routines **rrefstep**, which provides a step-by-step explanation, and **rowech**, which provides detailed checking of your work in the reduction process.

The reduced row echelon form of a matrix will be needed for a wide variety of future topics. The reduced row echelon form will be used to obtain information about the matrix itself which in turn will imply that the situation or problem modeled by the matrix will have certain properties. Hence we need a quick way to obtain the reduced row echelon form of a matrix $\boldsymbol{A}$ without supplying the detailed steps of the reduction process. For such purposes we use the command **rref**. Enter the matrix $\boldsymbol{A}$ given below into MATLAB , then command **rref(A)** displays its reduced row echelon form. Record the result next to matrix $\boldsymbol{A}$.

$$\boldsymbol{A} = \begin{bmatrix} 4 & 2 & -1 & 0 \\ 2 & -1 & 3 & 4 \\ 2 & 3 & -4 & -4 \\ 5 & -1 & 0 & 0 \end{bmatrix} \qquad \begin{bmatrix} \quad & \quad & \quad & \quad \\ & & & \\ & & & \\ & & & \end{bmatrix}$$

The **rref** command is so useful throughout this manual that we single out its format.

The command **rref(A)** produces the reduced row echelon form of a previously defined matrix $\boldsymbol{A}$.

The **rref** command automates the reduction steps that you can perform using routine **rowop**. The same relationship exists between routines **symrref** and **symrowop**; that is, **symrref** automatically produces the reduced row echelon form of a symbolic matrix $\boldsymbol{S}$. (See Section 2.3.) When used in the form

$$\textbf{[B R] = symrref(S)}$$

$\boldsymbol{B}$ will contain the rref of $\boldsymbol{S}$ while $\boldsymbol{R}$ is an *'informational matrix'* listing the quantities assumed not zero in the reduction process.[1] The information displayed in $\boldsymbol{R}$ can be used to gain insight into the solution set of a linear system. See Example 1.

[1]The restrictions displayed do not necessarily imply that the system has no solution if they are violated. Rather the restrictions listed were used along with a certain set of row operations to obtain the rref. Interchanging the order of the rows in the matrix could lead to a different set of restrictions used in the reduction process.

Example 1. Determine the values of parameter a for which the linear system

$$\begin{aligned}(3+a)x_1 + 2x_2 + 3x_3 &= 4\\ -x_1 + x_2 - x_3 &= -2\\ x_2 + ax_3 &= 0\end{aligned}$$

is inconsistent.

In MATLAB we form the symbolic matrix that represents the augmented matrix of the linear system and then use routine **symrref** to display the restrictions required to perform the reduction. The following commands perform the task and their results are shown as they appear in MATLAB .

```
S=sym('[3+a, 2, 3, 4;-1, 1, -1, -2;0, 1, a, 0]')



S =

[3+a,2, 3, 4]
[ -1,1,-1,-2]
[  0,1, a, 0]

[B R]=symrref(S)

B =

[1, 0, 0, 2*(4*a-1)/(6+a)/a]
[0, 1, 0,     -2*(1+a)/(6+a)]
[0, 0, 1,   2*(1+a)/(6+a)/a]

R =

[          RESTRICTIONS:]
[          The following]
[are assumed not zero.]
[                    3+a]
[            (5+a)/(3+a)]
[          (6+a)*a/(5+a)]
[                       ]
[                       ]
```

From the contents of matrix $\boldsymbol{R}$ above we see that it was assumed that

$$a \neq -3, \qquad a \neq -5, \qquad a \neq 0, \qquad a \neq -6$$

To determine which of these values lead to an inconsistent system we substitute the value into matrix $\boldsymbol{S}$, convert it to a numeric matrix, and determine the rref. The following commands show how to do this and display the result for the value $a = -3$.

```
P=subs(S,'a',-3)    %substituting -3 for a and naming the matrix P

P =

     0     2     3     4
    -1     1    -1    -2
     0     1    -3     0

rref(P)    %finding the rref

ans =

    1.0000         0         0    2.8889
         0    1.0000         0    1.3333
         0         0    1.0000    0.4444
```

The rref displayed above indicates that the linear system is consistent when $a = -3$. The above commands can be combined into a single command:

rref(subs(S, 'a', −3))

In this case only the rref will be displayed. To fully answer the question in this example we must investigate the behavior when a is -5, 0, and -6. For $a = -6$ we have

```
rref(subs(S,'a',-6))

ans =

     1     0    -5     0
     0     1    -6     0
     0     0     0     1
```

Hence the linear system is inconsistent for $a = -6$.

Enter the symbolic matrix $\boldsymbol{S}$ into MATLAB . Determine if the system is inconsistent for $a = -5$ and $a = 0$ using commands as illustrated above. Record your results below.

For $a = -5$ the system is _______________ . For $a = 0$ the system is _______________ .

Describe the set of values for a which make this system inconsistent:

____

Exercises 4.2

1. Use **rowop** to find the reduced row echelon form of each of the following matrices. Record the rref below each matrix.

$$C1 = \begin{bmatrix} 3 & -1 & 2 \\ 1 & 2 & 4 \\ -1 & 5 & 6 \\ 4 & 1 & 6 \end{bmatrix} \qquad C2 = \begin{bmatrix} 4 & 1 & 2 \\ -1 & 2 & 1 \\ 1 & 7 & 5 \end{bmatrix} \qquad C3 = \begin{bmatrix} 2 & 1 & 0 & 0 \\ 1 & 2 & 1 & 0 \\ 0 & 1 & 2 & 1 \\ 0 & 0 & 1 & 2 \end{bmatrix}$$

2. Use **rref** to find the general solution of the following homogeneous system of linear equations. Record your solution next to the linear system.

$$\begin{aligned} x_1 - x_2 + 2x_3 \qquad\quad + x_5 &= 0 \\ 2x_1 + x_2 + x_3 + x_4 + x_5 &= 0 \\ x_1 + x_2 \qquad\quad + 2x_4 + 2x_5 &= 0 \end{aligned}$$

3. Let $\boldsymbol{A}$ be the coefficient matrix in Exercise 2. Compute **rref(A)** and **rref(A′)**. Are they the same? Record both results in the space provided.

4. Construct a 4×4 matrix with two rows the same, but not all zeros. Compute its **rref**. Explain why there is a row of zeros in the rref.

5. Let $\boldsymbol{A}$ be an $n \times n$ matrix, $\boldsymbol{x}$ a column matrix of size $n \times 1$, and r a scalar. The matrix equation $\boldsymbol{Ax} = r\boldsymbol{x}$ arises in a variety of applications (see Lab 13). To solve this matrix equation we must determine both a column $\boldsymbol{x}$ and a scalar r. Here we show how to solve for $\boldsymbol{x}$ when r is known. Using matrix algebra we convert the matrix equation into an equivalent form so that we can solve a homogeneous linear system for $\boldsymbol{x}$.

$$\boldsymbol{Ax} = r\boldsymbol{x} \iff \boldsymbol{Ax} - r\boldsymbol{x} = \boldsymbol{0} \iff \boldsymbol{Ax} - r\boldsymbol{Ix} = \boldsymbol{0} \iff (\boldsymbol{A} - r\boldsymbol{I})\boldsymbol{x} = \boldsymbol{0}$$

This means that $\boldsymbol{x}$ is a solution of $\boldsymbol{A}\boldsymbol{x} = r\boldsymbol{x}$ if and only if $\boldsymbol{x}$ is a solution of the homogeneous system $(\boldsymbol{A} - r\boldsymbol{I})\boldsymbol{x} = \boldsymbol{0}$. Hence to find $\boldsymbol{x}$ we find the rref of $(\boldsymbol{A} - r\boldsymbol{I})$ and determine the general solution.

Find the general solution of the corresponding homogeneous system in each of the following using **rref**. Record your result below the expressions.

a) $\boldsymbol{A} = \begin{bmatrix} 1 & 2 \\ 2 & 1 \end{bmatrix}$ and $r = 3$. **b)** $\boldsymbol{A} = \begin{bmatrix} 0 & -2 & 1 \\ 1 & 3 & -1 \\ 0 & 0 & 1 \end{bmatrix}$ and $r = 1$.

6. In the population models discussed in Section 3.4, it can be argued that a steady state exists only if the matrix equation $\boldsymbol{AX} = \boldsymbol{X}$ has a nontrivial solution $\boldsymbol{X}$. Use the approach in Exercise 5 to determine the steady state when the transition matrix is

$$\boldsymbol{A} = \begin{bmatrix} .559 & .6 & .1 \\ .7 & 0 & 0 \\ 0 & .3 & 0 \end{bmatrix}$$

Record the steady state vector in the space provided.

7. Determine the values of parameter a for which the linear system

$$\begin{aligned} 2x_1 + x_2 + 3x_3 &= 7 \\ (a-1)x_1 + 4x_2 + 2x_3 &= 17 \\ x_2 + 2x_3 &= a \end{aligned}$$

is inconsistent. Describe your procedure and summarize the results.

8. Determine all values of parameter c for which the linear system
$$\begin{aligned} 3x_1 + \quad\quad\quad 12x_2 &= 0 \\ -2x_1 + (c^2 - 9)x_2 &= 0 \end{aligned}$$
has more than one solution.

$c =$ ______________________________

9. Determine all values of parameter c for which the linear system
$$\begin{aligned} x_1 + \ x_2 - \quad\quad\quad x_3 &= 2 \\ x_1 + 2x_2 + \quad\quad\quad x_3 &= 3 \\ x_1 + \ x_2 + (c^2 - 5)x_3 &= c \end{aligned}$$
has more than one solution.

$c =$ ______________________________

10. Quantities x_1, x_2, and x_3 are all non-negative and are related by the system of equations

$$\begin{aligned} x_1 + \ \ x_2 + \ \ 3x_3 &= 20000 \\ 50x_1 + 95x_2 + 145x_3 &= 1000 * m \\ -x_1 \quad\quad\quad + \ \ 2x_3 &= 0 \end{aligned}$$

Determine the largest value that parameter m can have so that the system is consistent. Describe your procedure and summarize the results in the space below.

Section 4.3

Inverses

To determine if an $n \times n$ matrix $\boldsymbol{A}$ has an inverse, just apply **rref** to the matrix $[\boldsymbol{A}|\boldsymbol{I_n}]$. Enter the matrix $\boldsymbol{A}$ into MATLAB and use the command

rref([A eye(size(A))])

If $\boldsymbol{A}$ is transformed into $\boldsymbol{I_n}$, then $\boldsymbol{I_n}$ will have been transformed into $\boldsymbol{A}^{-1}$; otherwise $\boldsymbol{A}$ is singular (that is, not invertible). For practice find the inverse of each of the following:

$$\mathbf{A1} = \begin{bmatrix} 1 & 2 & 0 \\ 0 & 1 & -1 \\ 1 & 0 & 1 \end{bmatrix}, \quad \mathbf{A2} = \begin{bmatrix} 1 & 1 & 0 \\ 0 & 1 & 1 \\ 1 & 0 & 1 \end{bmatrix}$$

Record your results here:

$$\text{inverse of } \mathbf{A1} = \begin{bmatrix} \quad & \quad & \quad \\ & & \\ & & \end{bmatrix}, \quad \text{inverse of } \mathbf{A2} = \begin{bmatrix} \quad & \quad & \quad \\ & & \\ & & \end{bmatrix}$$

The exercises contain two other ways to compute matrix inverses in MATLAB . Carefully note the MATLAB commands for future use. The results contained in Exercise 6 are especially revealing.

Exercises 4.3

1. Use the **rref** command to find the inverse of each matrix below. Put the results in the space provided.

$$\mathbf{A1} = \begin{bmatrix} 1 & 1 & 0 \\ 2 & 0 & 1 \\ 1 & 0 & 1 \end{bmatrix} \qquad \mathbf{A2} = \begin{bmatrix} 1 & 1 & 0 \\ 0 & 1 & 1 \\ 1 & 0 & 1 \end{bmatrix}$$

2. There are several routines for computing the inverse of a matrix. One of these is the command **invert**. Type **help invert** for directions. This command employs **rref** and just avoids the requirement of appending the same size identity matrix. Use **invert** to determine which of the following matrices are nonsingular. Beside each matrix record if it is singular or nonsingular. Record the inverse if it is nonsingular.

a) $A = \begin{bmatrix} 1 & 2 & 3 \\ 4 & 5 & 6 \\ 7 & 8 & 9 \end{bmatrix}$ b) $B = \begin{bmatrix} 1 & 2 & 3 \\ 4 & 5 & 6 \\ 7 & 8 & 0 \end{bmatrix}$

c) $C = \begin{bmatrix} 1 & 2 & 3 & 0 \\ 4 & 5 & 0 & 6 \\ 7 & 0 & 8 & 9 \\ 0 & 10 & 11 & 12 \end{bmatrix}$ d) $D = \begin{bmatrix} 1 & 2 & 3 & 0 \\ 4 & 5 & 0 & 6 \\ 7 & 0 & 8 & 9 \\ 1 & 2 & 3 & 0 \end{bmatrix}$

3. The command **invert** in Exercise 2 is not as efficient as it could be. MATLAB command **inv** uses a different strategy for computing inverses that we do not study in detail. Find the inverses of the matrices in Exercise 2 using **inv** and compare the results with those of Exercise 2. (Warning: MATLAB does not use exact arithmetic and for some matrices **inv** may display a message indicating potential inaccuracies may be present.)

4. If $\boldsymbol{B}$ is the inverse of an $n \times n$ matrix $\boldsymbol{A}$ then $\boldsymbol{AB} = \boldsymbol{BA} = \boldsymbol{I_n}$ assuming that exact arithmetic is used for the computations. If computer arithmetic is used to compute the product $\boldsymbol{AB}$, then the result need not equal $\boldsymbol{I_n}$ and in fact $\boldsymbol{AB}$ need not equal $\boldsymbol{BA}$. However, both $\boldsymbol{AB}$ and $\boldsymbol{BA}$ should be close to $\boldsymbol{I_n}$. Use the **invert** command from Exercise 2 in the form **B = invert(A)**, then compute the products $\boldsymbol{AB}$ and $\boldsymbol{BA}$ in MATLAB for the following matrices.

a) $\begin{bmatrix} 1 & \frac{1}{3} \\ 0 & \frac{1}{3} \end{bmatrix}$ b) $\begin{bmatrix} \frac{1}{2} & \frac{1}{4} \\ \frac{1}{4} & \frac{1}{2} \end{bmatrix}$ c) $\begin{bmatrix} 1 & \frac{1}{2} & \frac{1}{3} \\ \frac{1}{2} & \frac{1}{3} & \frac{1}{4} \\ \frac{1}{3} & \frac{1}{4} & \frac{1}{5} \end{bmatrix}$

5. If $\boldsymbol{A}$ is a nonsingular (invertible) matrix the solution of the linear system $\boldsymbol{Ax} = \boldsymbol{c}$ is $\boldsymbol{x} = \boldsymbol{A}^{-1}\boldsymbol{c}$. Solve each of the systems below by TWO methods: first using **rowop** and second using the **invert** command as described above. Compare the two solutions; are they the same? Briefly describe any differences in the space below.

a) $A = \begin{bmatrix} \frac{1}{2} & \frac{2}{3} & \frac{4}{3} \\ 0 & \frac{2}{3} & \frac{4}{3} \\ 0 & 0 & \frac{5}{3} \end{bmatrix}$, $c = \begin{bmatrix} 2 \\ 2 \\ \frac{10}{3} \end{bmatrix}$

b) $\boldsymbol{A} = [a_{ij}]$, where $a_{ij} = \dfrac{1}{i+j-1}$ for i, j = 1, 2,...,10 and $\boldsymbol{c}$ is the first column of

I_{10}. To generate $\boldsymbol{A}$ use MATLAB commands
for i=1:10,for j=1:10,A(i,j)=1/(i+j-1);end,end,A

6. Let $\boldsymbol{A}$ be a nonsingular matrix. The linear system $\boldsymbol{Ax} = \boldsymbol{b}$ can be solved in MATLAB in several ways. (Assume $\boldsymbol{A}$ and $\boldsymbol{b}$ have been entered into MATLAB .)

(i) **rref([A b])** becomes **[I x]**.

(ii) Compute $\boldsymbol{A}^{-1}$, then $\boldsymbol{x} = \boldsymbol{A}^{-1}\boldsymbol{b}$. In MATLAB we use command

x = invert(A)*b or **x = inv(A)*b**

(iii) MATLAB is designed to solve linear systems of equations. Hence it has a special command for this purpose. The command is \ which invokes an algorithm other than **rref** or **inv**. To use this command enter **x = A\b**.

One way to compare these methods is to count the number of arithmetic operations they require. In MATLAB we call the arithmetic operations *floating point operations.* MATLAB has a command, **flops**, which counts such operations. (For more information use **help**.) In the following comparisons we use the Hilbert matrix (see command **hilb**) for the coefficient matrix $\boldsymbol{A}$ and a column of all ones for $\boldsymbol{b}$. Use the following commands, changing n to the appropriate value for the matrix size, to get flop counts for the three methods. Record the flop counts **num1, num2**, and **num3** in the table.

```
n = 5;A = hilb(n);b=ones(n,1);
flops(0);rref([A b]);num1 = flops
flops(0);x = inv(A)*b;num2 = flops
flops(0);x = A\b;num3 = flops
```

n	rref	inv	\
5			
10			
25			

You should find that using \ is cheaper (fewer flops) hence faster and in numerical analysis courses it is shown that \ usually gives more accurate results. The message from the preceding table should be clear: **to solve nonsingular linear systems use \.**

7. Determine the values of the parameter k that makes the following matrix invertible.

$$\boldsymbol{A} = \begin{bmatrix} 1 & 1 & 2 \\ 2-k & 0 & 1 \\ 0 & 5 & k+12 \end{bmatrix}$$

Describe your procedure and summarize the results in the space below. (Hint: use **sym-rref**.)

8. Find all values of parameters c and d which make the matrix $\begin{bmatrix} 5 & c \\ 5 & d \end{bmatrix}$ invertible.

9. Find all values of parameters c which make the matrix $\begin{bmatrix} c & c+1 & c+2 \\ c+2 & c & c+1 \\ c+1 & c+2 & c \end{bmatrix}$ invertible.

LAB 5

A Vector Space Example

Topics: vector space axioms applied to a particular set V with operations $\oplus$ and $\odot$.

Introduction

A vector space V is a set of objects, called vectors, with two operations. It is standard to use the name 'addition' for one operation and to denote it by $\oplus$. It is also standard to use the name 'scalar multiplication' for the other operation and to denote it by $\odot$.

One purpose of this lab is to show that caution must be exercised when using these terms and symbols. Another purpose is to clarify several parts of the definition of a vector space that tend to get overlooked when only ordinary operations of matrix addition and scalar multiplication are used.

There are ten properties in the definition of a vector space. We will examine each property separately for the examples considered here. You can use MATLAB to help you determine whether a given property is true by experimenting with several different vectors and scalars. If you find a vector or pair of vectors for which a property does not hold then that property is not true. However, to show that a property is true you must use a general written argument that the property holds for **all** vectors in V and, if they are needed, for **all** real scalars. The box below summarizes this discussion.

> MATLAB can only be used to show that a property is NOT true by generating particular matrices for which the property does not hold. To show that a property is TRUE you must supply an argument developed by hand that holds for all vectors and, if necessary, for all real scalars.

Section 5.1

Experimenting with Vector Space Properties

For this section the set V will consist of all 3×3 real matrices.

> V = all 3×3 real matrices

You can randomly generate a 3×3 matrix $\boldsymbol{A}$ with integer entries by typing the following command in MATLAB :

$$\mathbf{A} = \mathbf{fix}(10{*}\mathbf{randn}(3))$$

The next time you want another 3×3 matrix use the same command but name the matrix something other than $\boldsymbol{A}$.

The operations on the set V will be defined in terms of MATLAB operations. Addition is defined by

$$\boxed{\boldsymbol{A} \oplus \boldsymbol{B} = \mathbf{A.*B}}$$

Notice that this operation has two parts, a period and an asterisk. This is 'addition' in V, even though it resembles multiplication.

Now generate two 3×3 matrices $\boldsymbol{A}$ and $\boldsymbol{B}$ in MATLAB . Then type $\boldsymbol{A}.*\boldsymbol{B}$ to determine the action that takes place. Do this enough times on different pairs of matrices so you can describe the operation $\oplus$ in words. Record your description of this 'addition' operation here.

____

____

There are five properties in the definition of a vector space that deal with $\oplus$ alone. We will examine each property separately. If you feel that property (1), (2), or (3) holds you must show that it is true for arbitrary matrices by dealing with general entries. (That is, you can not use matrices with particular numerical entries to verify that the property is true.) If you feel that it is not true then you must supply specific matrices and verify that the property fails on these matrices.

(1) The operation $\oplus$ is **closed** on V if $\boldsymbol{A} \oplus \boldsymbol{B}$ is in V for all 3×3 matrices $\boldsymbol{A}$ and $\boldsymbol{B}$.

(2) The operation $\oplus$ is **commutative** on V if $\boxed{\boldsymbol{A} \oplus \boldsymbol{B} = \boldsymbol{B} \oplus \boldsymbol{A}}$ for all 3×3 matrices $\boldsymbol{A}$ and $\boldsymbol{B}$.

(3) The operation $\oplus$ is **associative** on V if $\boxed{(\mathbf{A} \oplus \boldsymbol{B}) \oplus \mathbf{C} = \boldsymbol{A} \oplus (\mathbf{B} \ \oplus \mathbf{C})}$ for all 3×3 matrices $\boldsymbol{A}$, $\boldsymbol{B}$, and $\mathbf{C}$.

(4) V has the zero property if there is a 3×3 matrix $\mathbf{Z}$ such that $\boxed{\boldsymbol{A} \oplus \mathbf{Z} = \boldsymbol{A}}$ for each 3×3 matrix $\boldsymbol{A}$. If V has this property then you must list nine specific entries for the matrix $\mathbf{Z}$ which will work with every matrix $\boldsymbol{A}$.

(5) V has the inverse property if for each 3×3 matrix $\boldsymbol{A}$ there is a 3×3 matrix $\boldsymbol{X}$ such that $\boxed{\boldsymbol{A} \oplus \boldsymbol{X} = \mathbf{Z}}$, where $\mathbf{Z}$ is the matrix found in (4). If V has this property then you must begin with a general 3×3 matrix $\boldsymbol{A}$, then list nine specific entries for the matrix $\boldsymbol{X}$ so that $\boldsymbol{A} \oplus \boldsymbol{X} = \mathbf{Z}$.

The operation of scalar multiplication is defined by

$$\boxed{\mathrm{c} \odot \boldsymbol{A} = \mathrm{c} + \boldsymbol{A}}$$

Notice that 'scalar multiplication' looks like addition here.

Now generate a 3×3 matrix $\boldsymbol{A}$ and choose any scalar c. Then type c + **A** to determine the action that takes place. Do this enough times on different matrices and scalars so you can describe the operation $\odot$ in words. Record your description of this 'scalar multiplication' operation here.

---

(6) There is one property in the definition of a vector space that deals with $\odot$ alone. The operation $\odot$ is **closed** on V if c $\odot$ $\boldsymbol{A}$ is in V for all 3×3 matrices $\boldsymbol{A}$ and all real scalars c. If you feel that V is closed you must show that it is true for an arbitrary matrix **A** and an arbitrary real scalar c. If you feel that it is not true then you must supply a specific matrix and a specific scalar where c $\odot$ $\boldsymbol{A}$ is not in V.

There are two properties in the definition of a vector space that tie together the two operations $\oplus$ and $\odot$. Both properties look like the familiar distributive laws.

(7) This property is

$$\boxed{\mathrm{c} \odot (\boldsymbol{A} \oplus \boldsymbol{B}) = (\mathrm{c} \odot \boldsymbol{A}) \oplus (\mathrm{c} \odot \boldsymbol{B})}$$

The MATLAB version becomes c+($\boldsymbol{A}$.*$\boldsymbol{B}$) = (c+$\boldsymbol{A}$).*(c+$\boldsymbol{B}$). If you feel that this property holds you must show that it is true for arbitrary matrices $\boldsymbol{A}$ and $\boldsymbol{B}$ and an arbitrary scalar c. If you feel that it is not true then you must supply specific matrices and a specific scalar and verify that it fails.

(8) This property is

$$\boxed{(\mathrm{c} + \mathrm{d}) \odot \boldsymbol{A} = (\mathrm{c} \odot \boldsymbol{A}) \oplus (\mathrm{d} \odot \boldsymbol{A})}$$

Notice that the first operation on the left-hand side of this property is ordinary addition of real numbers. The MATLAB version becomes (c+d)+***A*** = (c+***A***).*(d+***A***). If you feel that this property holds you must show that it is true for an arbitrary matrix ***A*** and arbitrary scalars c and d. If you feel that it is not true then you must supply a specific matrix and specific scalars c and d, and verify that it fails.

(9) This property ties together ordinary multiplication of real numbers with the operation $\odot$:

$$\boxed{(\mathrm{cd}) \odot \boldsymbol{A} = \mathrm{c} \odot (\mathrm{d} \odot \mathbf{A})}$$

The first operation on the left-hand side is ordinary multiplication of real numbers. If you feel that this property holds you must show that it is true for an arbitrary matrix ***A*** and arbitrary scalars c and d. If you feel that it is not true then you must supply a specific matrix and specific scalars c and d, and verify that it fails.

(10) This final property concerns the scalar 1.

$$\boxed{1 \odot \boldsymbol{A} = \boldsymbol{A}}$$

If you feel that this property holds you must show that it is true for an arbitrary matrix ***A***. If you feel that it is not true then you must supply a specific matrix ***A*** and verify that it fails.

Is V with $\oplus$ and $\odot$ a vector space?	YES	NO	(Circle one.)

Section 5.2

More Experiments with Vector Space Properties

For this section the set $\boldsymbol{V}$ will consist of all 2×1 matrices with positive entries. The operations on the set $\boldsymbol{V}$ will be defined in terms of MATLAB operations. Addition is defined by

$$\boxed{\boldsymbol{v} \oplus \boldsymbol{w} = \boldsymbol{v}. * \boldsymbol{w}}$$

where $\boldsymbol{v}$ and $\boldsymbol{w}$ are in $\boldsymbol{V}$. Notice that this operation has two parts, a period and an asterisk. This MATLAB operation performs entry-by-entry multiplication, but we use this as our addition operation.

Example 1. For $\boldsymbol{v} = \begin{bmatrix} 3 \\ 7 \end{bmatrix}$ and $\boldsymbol{w} = \begin{bmatrix} 5 \\ 1/2 \end{bmatrix}$ we have

$$\boldsymbol{v} \oplus \boldsymbol{w} = \boldsymbol{v}.{*}\boldsymbol{w} = \begin{bmatrix} (3)(5) \\ (7)(1/2) \end{bmatrix} = \begin{bmatrix} 15 \\ 7/2 \end{bmatrix}$$

Scalar multiplication on $\boldsymbol{V}$ is defined by

$$\boxed{k \odot \boldsymbol{v} = \boldsymbol{v}.\hat{}\,k}$$

where $\boldsymbol{v}$ is in $\boldsymbol{V}$ and k is a real scalar. Notice that this operation also has two parts, a period and the exponentiation symbol, a carat. This MATLAB operation performs entry-by-entry exponentiation of the entries in $\boldsymbol{v}$ to the kth power.

Example 2. For $k = 1/2$ and $\boldsymbol{v} = \begin{bmatrix} 16 \\ 3 \end{bmatrix}$, we have

$$k \odot \boldsymbol{v} = \boldsymbol{v}.\hat{}\,k = \begin{bmatrix} 16^{1/2} \\ 3^{1/2} \end{bmatrix} \approx \begin{bmatrix} 4 \\ 1.732 \end{bmatrix}$$

The question is,

"Is $\boldsymbol{V}$ with the operations $\oplus$ and $\odot$ as defined above, a vector space?"

There are a total of ten properties to be checked. Numerical experiments with the operations of $\oplus$ and $\odot$ can provide you with evidence that a property may be valid. However to claim that a property is true you must verify the result for all vectors $\boldsymbol{v}$ in $\boldsymbol{V}$ and all real scalars k. In order to show that a property is invalid, you can show that the result is not true for specific vectors in $\boldsymbol{V}$ or real scalars.

For each of the following circle the appropriate response. Supply evidence that supports your choice.

1. $\boldsymbol{V}$ is closed under addition using $\oplus$ True False

2. $\oplus$ is commutative on $\boldsymbol{V}$. True False

3. $\oplus$ is associative on $\boldsymbol{V}$. True False

4. $\boldsymbol{V}$ has a "zero vector"; that is, there is a vector $\boldsymbol{z}$ in $\boldsymbol{V}$ so that for all $\boldsymbol{v}$ in $\boldsymbol{V}$ $\boxed{\boldsymbol{v} \oplus \boldsymbol{z} = \boldsymbol{v}}$.

True False

5. $\boldsymbol{V}$ has an "inverse property"; that is, for a vector $\boldsymbol{v}$ in $\boldsymbol{V}$ there is a vector $\boldsymbol{x}$ in $\boldsymbol{V}$ so that, $\boxed{\boldsymbol{v} \oplus \boldsymbol{x} = \boldsymbol{z}}$ where $\boldsymbol{z}$ is the "zero vector" from (4).

True False

6. $\boldsymbol{V}$ is closed under $\odot$. True False

7. $\boxed{k \odot (\boldsymbol{v} \oplus \boldsymbol{w}) = (k \odot \boldsymbol{v}) \oplus (k \odot \boldsymbol{w})}$ for all $\boldsymbol{v}$ and $\boldsymbol{w}$ in $\boldsymbol{V}$ and any real scalar k

True False

8. $\boxed{(k + t) \odot \boldsymbol{v} = (k \odot \boldsymbol{v}) \oplus (t \odot \boldsymbol{v})}$ for all $\boldsymbol{v}$ in $\boldsymbol{V}$ and any real scalars k and t

True False

9. $\boxed{(kt) \odot \boldsymbol{v} = k \odot (t \odot \boldsymbol{v})}$ for all $\boldsymbol{v}$ and $\boldsymbol{w}$ in $\boldsymbol{V}$ and any real scalars k and t

True False

10. For the scalar 1, $\boxed{1 \odot \boldsymbol{v} = \boldsymbol{v}}$ for any vector $\boldsymbol{v}$ in $\boldsymbol{V}$. True False

Is $\boldsymbol{V}$ with $\oplus$ and $\odot$ a vector space? Yes No (Circle one)

<< NOTES; COMMENTS; IDEAS >>

LAB 6

Linear Combinations

Topics: linear combinations; span; linear independence and linear dependence; basis.

Introduction

The notion of a linear combination is fundamental to a wide variety of topics in linear algebra. The ideas of span, linear independence, linear dependence, and basis center on forming linear combinations of vectors. In addition, elementary row operations are essentially of the form 'replace an existing row by a linear combination of rows'. This is clearly the case when we add a multiple of one row to another row. From this point of view it follows that the reduced row echelon form and the row echelon form are processes for implementing linear combinations of **rows** of a matrix. Hence the MATLAB routines **rowop** and **rref** should be useful in solving problems that involve linear combinations.

Here we discuss how to use MATLAB to solve problems dealing with linear combinations (Section 6.1), span (Section 6.2), linear independence, linear dependence (Section 6.3), and basis (Section 6.4). The basic strategy is to set up a linear system related to the problem and ask questions like,

'Is there a solution?'

'Is the only solution the trivial solution?'

It is important to recognize that the technique for deriving the appropriate linear system will vary according to the type of vectors involved.

Sections 2.1 and 4.1 are essential for this entire lab. In addition, Section 6.5, which makes use of routine **symrowop** to illustrate how the basic concepts can be handled when the augmented matrix is symbolic, depends on Section 2.3.

Section 6.1

Linear Combinations

Given a vector space $\mathbf{V}$ and a set of vectors $\mathbf{S} = \{ \boldsymbol{X_1}, \boldsymbol{X_2}, ..., \boldsymbol{X_k} \}$ in $\mathbf{V}$, determine if $\boldsymbol{X}$, belonging to $\mathbf{V}$, can be expressed as a linear combination of the members of $\mathbf{S}$. That is, can we find some set of scalars $c_1, c_2, ..., c_k$ so that

$$c_1\boldsymbol{X_1} + c_2\boldsymbol{X_2} + \cdots + c_k\boldsymbol{X_k} = \boldsymbol{X}$$

There are several common situations which we illustrate by example and then provide a summary of the method employed.

Example 1. Let $\boldsymbol{X_1} = (1, 2, 1, -1)$, $\boldsymbol{X_2} = (1, 0, 2, -3)$, and $\boldsymbol{X_3} = (1, 1, 0, -2)$. Determine if the vector $\boldsymbol{X} = (2, 1, 5, -5)$ is a linear combination of $\boldsymbol{X_1}$, $\boldsymbol{X_2}$, and $\boldsymbol{X_3}$.

Form the expression

$$\sum_{j=1}^{3} c_j \boldsymbol{X_j} = c_1\boldsymbol{X_1} + c_2\boldsymbol{X_2} + c_3\boldsymbol{X_3} = \boldsymbol{X}$$

and find the corresponding linear system to solve for the c_i. We have

$$c_1(1,2,1,-1) + c_2(1,0,2,-3) + c_3(1,1,0,-2) = (2,1,5,-5).$$

Performing scalar multiplications and adding corresponding entries gives

$$(c_1 + c_2 + c_3, 2c_1 + c_3, c_1 + 2c_2, -c_1 - 3c_2 - 2c_3) = (2,1,5,-5)$$

Since the two vectors are equal, we equate corresponding entries to obtain the system of equations

$$\begin{aligned} c_1 + c_2 + c_3 &= 2 \\ 2c_1 \quad\quad + c_3 &= 1 \\ c_1 + 2c_2 \quad\quad &= 5 \\ -c_1 - 3c_2 - 2c_3 &= -5 \end{aligned}$$

The matrix form for this linear system is $\boldsymbol{Ac} = \boldsymbol{b}$ where

$$\boldsymbol{A} = \begin{bmatrix} 1 & 1 & 1 \\ 2 & 0 & 1 \\ 1 & 2 & 0 \\ -1 & -3 & -2 \end{bmatrix}, \ \boldsymbol{c} = \begin{bmatrix} c_1 \\ c_2 \\ c_3 \end{bmatrix}, \text{ and } \boldsymbol{b} = \begin{bmatrix} 2 \\ 1 \\ 5 \\ -5 \end{bmatrix}$$

Note: the columns of $\boldsymbol{A}$ are the transposes of the original vectors.

Enter $\boldsymbol{A}$ and $\boldsymbol{b}$ into MATLAB and then use command

rref([A b])

to give

```
ans =

     1     0     0     1
     0     1     0     2
     0     0     1    -1
     0     0     0     0
```

Recall that this display represents the reduced row echelon form of an augmented matrix. It follows that the system is consistent with solution

$$c_1 = 1, \quad c_2 = 2, \quad c_3 = -1$$

Hence $\boldsymbol{X}$ is a linear combination of $\boldsymbol{X_1}$, $\boldsymbol{X_2}$, and $\boldsymbol{X_3}$, with

$$\boldsymbol{X_1} + 2\boldsymbol{X_2} - \boldsymbol{X_3} = \boldsymbol{X}\,.$$

I. Strategy summary for linear combinations of rows.

If the vectors $\boldsymbol{X}_i$ in **S** are row matrices, then we construct a linear system whose coefficient matrix $\boldsymbol{A}$ is

$$\boldsymbol{A} = \begin{bmatrix} \boldsymbol{X_1} \\ \boldsymbol{X_2} \\ \vdots \\ \boldsymbol{X_k} \end{bmatrix}^T = \begin{bmatrix} \boldsymbol{X_1^T} & \boldsymbol{X_2^T} & \cdots & \boldsymbol{X_k^T} \end{bmatrix}$$

and whose right-hand side is $\boldsymbol{X^T}$. That is, the columns of **A** are the row matrices of set **S** converted to columns. Let $\boldsymbol{c} = \begin{bmatrix} c_1 & c_2 & \cdots & c_k \end{bmatrix}^T$ and $\boldsymbol{b} = \boldsymbol{X^T}$, then try to solve the linear system $\boldsymbol{Ac} = \boldsymbol{b}$ using **rowop** or **rref** in MATLAB . If the system is shown to be consistent, having no rows of the form $\left[\begin{array}{cccc|c} 0 & 0 & \ldots & 0 & q \end{array}\right]$, $q \neq 0$, then the vector $\boldsymbol{X}$ can be written as a linear combination of the vectors in **S**. In that case the solution of the system gives the values of the coefficients. Caution: Many times we need only determine if the system is consistent to answer that $\boldsymbol{X}$ is a linear combination of the members of **S**.

Example 2. Let $\boldsymbol{X_1} = \begin{bmatrix} 1 \\ 2 \\ 1 \end{bmatrix}$, $\boldsymbol{X_2} = \begin{bmatrix} -1 \\ 1 \\ 3 \end{bmatrix}$, $\boldsymbol{X_3} = \begin{bmatrix} 1 \\ 5 \\ 5 \end{bmatrix}$, $\boldsymbol{X_4} = \begin{bmatrix} 3 \\ 0 \\ -5 \end{bmatrix}$. Determine if $\boldsymbol{X} = \begin{bmatrix} 1 \\ 0 \\ 2 \end{bmatrix}$ is a linear combination of $\boldsymbol{X_1}, \boldsymbol{X_2}, \boldsymbol{X_3}, \boldsymbol{X_4}$.

Form the expression

$$\sum_{j=1}^{4} c_j \boldsymbol{X_j} = c_1\boldsymbol{X_1} + c_2\boldsymbol{X_2} + c_3\boldsymbol{X_3} + c_4\boldsymbol{X_4} = \boldsymbol{X}$$

and find the corresponding linear system to solve for the c_i. We have

$$c_1 \begin{bmatrix} 1 \\ 2 \\ 1 \end{bmatrix} + c_2 \begin{bmatrix} -1 \\ 1 \\ 3 \end{bmatrix} + c_3 \begin{bmatrix} 1 \\ 5 \\ 5 \end{bmatrix} + c_4 \begin{bmatrix} 3 \\ 0 \\ -5 \end{bmatrix} = \begin{bmatrix} 1 \\ 0 \\ 2 \end{bmatrix}$$

Performing the scalar multiplications and adding corresponding entries gives

$$\begin{bmatrix} c_1 - c_2 + c_3 + 3c_4 \\ 2c_1 + c_2 + 5c_3 \\ c_1 + 3c_2 + 5c_3 - 5c_4 \end{bmatrix} = \begin{bmatrix} 1 \\ 0 \\ 2 \end{bmatrix}$$

Since the two vectors are equal, we equate corresponding entries to obtain the system of equations

$$\begin{aligned} c_1 - c_2 + c_3 + 3c_4 &= 1 \\ 2c_1 + c_2 + 5c_3 \phantom{{}+3c_4} &= 0 \\ c_1 + 3c_2 + 5c_3 - 5c_4 &= 2 \end{aligned}$$

The matrix form for this linear system is $\boldsymbol{Ac} = \boldsymbol{b}$ where

$$\boldsymbol{A} = \begin{bmatrix} 1 & -1 & 1 & 3 \\ 2 & 1 & 5 & 0 \\ 1 & 3 & 5 & -5 \end{bmatrix}, \ \boldsymbol{c} = \begin{bmatrix} c_1 \\ c_2 \\ c_3 \\ c_4 \end{bmatrix}, \text{ and } \boldsymbol{b} = \begin{bmatrix} 1 \\ 0 \\ 2 \end{bmatrix}$$

Note: the columns of $\boldsymbol{A}$ are just the original columns.

Enter $\boldsymbol{A}$ and $\boldsymbol{b}$ into MATLAB and then use command

rref([A b])

to give

```
ans =

     1     0     2     1     0
     0     1     1    -2     0
     0     0     0     0     1
```

Recall that this display represents the reduced row echelon form of an augmented matrix. The bottom row indicates that the system is inconsistent. Hence $\boldsymbol{X}$ is not a linear combination of $\boldsymbol{X_1}$, $\boldsymbol{X_2}$, $\boldsymbol{X_3}$, and $\boldsymbol{X_4}$.

II. Strategy summary for linear combinations of columns.

If the vectors $\boldsymbol{X}_i$ in **S** are column matrices, then just lay the columns side-by-side to form the coefficient matrix

$$\boldsymbol{A} = \begin{bmatrix} \boldsymbol{X}_1 & \boldsymbol{X}_2 & \cdots & \boldsymbol{X}_k \end{bmatrix}$$

and set $\boldsymbol{b} = \boldsymbol{X}$. Proceed as described in **I**.

III. Strategy summary for linear combinations of polynomials.

If the vectors in **S** are polynomials, then associate with each polynomial a column of coefficients. Make sure any missing terms in the polynomial are associated with a zero coefficient. One way to proceed is to use the coefficient of the highest powered term as the first entry of the column, the coefficient of the next highest powered term second, and so on. For example,

$$t^2 + 2t + 1 \longrightarrow \begin{bmatrix} 1 \\ 2 \\ 1 \end{bmatrix} \qquad t^2 + 2 \longrightarrow \begin{bmatrix} 1 \\ 0 \\ 2 \end{bmatrix} \qquad 3t - 2 \longrightarrow \begin{bmatrix} 0 \\ 3 \\ -2 \end{bmatrix}$$

The linear combination problem is now solved as in II.

Example 3. Given matrix $\boldsymbol{P} = \begin{bmatrix} 1 & 2 & 3 \\ 4 & 5 & 6 \end{bmatrix}$. To associate a column matrix as described above within MATLAB , first enter $\boldsymbol{P}$ into MATLAB , then type command

X = reshape(P,6,1)

which gives

```
X =

     1
     4
     2
     5
     3
     6
```

For more information type **help reshape**.

IV. Strategy summary for linear combinations of $m \times n$ matrices.

If the vectors in **S** are $m \times n$ matrices, then associate with each such matrix $\boldsymbol{A_j}$ a column $\boldsymbol{X_j}$ formed by stringing together its columns one below the other from left to right. In MATLAB this transformation is done using the **reshape** command. Then we proceed as in II.

Example 4. Let $\boldsymbol{M}1 = \begin{bmatrix} -1 & 0 & 1 \\ 4 & 1 & 0 \end{bmatrix}$, $\boldsymbol{M}2 = \begin{bmatrix} 6 & 3 & 0 \\ 1 & 2 & 0 \end{bmatrix}$, $\boldsymbol{M}3 = \begin{bmatrix} -7 & 8 & 14 \\ 26 & -11 & 17 \end{bmatrix}$, $\boldsymbol{M}4 = \begin{bmatrix} -1 & -2 & 0 \\ -3 & 4 & -1 \end{bmatrix}$. Determine if $\boldsymbol{M} = \begin{bmatrix} -14 & -4 & 3 \\ 13 & -5 & 1 \end{bmatrix}$ is a linear combination of $\boldsymbol{M}1$, $\boldsymbol{M}2$, $\boldsymbol{M}3$, $\boldsymbol{M}4$.

After entering the matrices, type commands

$$\mathbf{X1} = \mathbf{reshape(M1, 6, 1);} \quad \mathbf{X2} = \mathbf{reshape(M2, 6, 1);}$$
$$\mathbf{X3} = \mathbf{reshape(M3, 6, 1);} \quad \mathbf{X4} = \mathbf{reshape(M4, 6, 1);}$$
$$\mathbf{X} = \mathbf{reshape(M, 6, 1);}$$

Then command **rref([X1 X2 X3 X4 X])** gives

```
ans =

     1     0     0     0     3
     0     1     0     0    -2
     0     0     1     0     0
     0     0     0     1    -1
     0     0     0     0     0
     0     0     0     0     0
```

Hence $\boldsymbol{M}$ is a linear combination of $\boldsymbol{M}1$, $\boldsymbol{M}2$, $\boldsymbol{M}3$, $\boldsymbol{M}4$. In fact we have

$$3*\boldsymbol{M}1 - 2*\boldsymbol{M}2 - \boldsymbol{M}4 = \boldsymbol{M}$$

Exercises 6.1

1. Let $\boldsymbol{v_1} = [4 \quad 2 \quad 1]$, $\boldsymbol{v_2} = [-2 \quad 3 \quad 1]$, and $\boldsymbol{v_3} = [2 \quad -11 \quad -4]$. Determine if each of the following vectors $\boldsymbol{u}$ is a linear combination of $\boldsymbol{v_1}, \boldsymbol{v_2}$, and $\boldsymbol{v_3}$. If it is, then display the linear combination by supplying the coefficients and appropriate operations.

a) $\boldsymbol{u} = [6 \quad 5 \quad 5]$ Circle one: Yes No $\boldsymbol{u} =$ ___$\boldsymbol{v_1}$___$\boldsymbol{v_2}$___$\boldsymbol{v_3}$

b) $\boldsymbol{u} = [10 \quad -15 \quad -5]$ Circle one: Yes No $\boldsymbol{u} =$ ___$\boldsymbol{v_1}$___$\boldsymbol{v_2}$___$\boldsymbol{v_3}$

c) $\boldsymbol{u} = [9 \quad -17.5 \quad -6]$ Circle one: Yes No $\boldsymbol{u} =$ ___$\boldsymbol{v_1}$___$\boldsymbol{v_2}$___$\boldsymbol{v_3}$

2. Let $\boldsymbol{v_1} = \begin{bmatrix} 1 \\ -1 \\ 2 \\ 4 \end{bmatrix}$, $\boldsymbol{v_2} = \begin{bmatrix} 0 \\ 2 \\ 1 \\ 1 \end{bmatrix}$, and $\boldsymbol{v_3} = \begin{bmatrix} 3 \\ 1 \\ 0 \\ 2 \end{bmatrix}$. Determine if each of the following vectors $\boldsymbol{u}$ is a linear combination of $\boldsymbol{v_1}, \boldsymbol{v_2}$, and $\boldsymbol{v_3}$. If it is, then display the linear combination by supplying the coefficients and appropriate operations.

a) $\boldsymbol{u} = \begin{bmatrix} 11 \\ -1 \\ 3 \\ 13 \end{bmatrix}$ Circle one: Yes No $\boldsymbol{u} = \_\_\boldsymbol{v_1}\_\_\boldsymbol{v_2}\_\_\boldsymbol{v_3}$

b) $\boldsymbol{u} = \begin{bmatrix} 1 \\ 0 \\ 1 \\ 1 \end{bmatrix}$ Circle one: Yes No $\boldsymbol{u} = \_\_\boldsymbol{v_1}\_\_\boldsymbol{v_2}\_\_\boldsymbol{v_3}$

3. Let $p_1(t) = 2t^3 + t - 1$, $p_2(t) = t^2 + 2t$, and $p_3(t) = t^3 - t^2 + 3t + 2$. Determine if each of the following vectors $q(t)$ is a linear combination of $p_1(t), p_2(t)$, and $p_3(t)$. If it is, then display the linear combination by supplying the coefficients and appropriate operations.

a) $q(t) = 4t^2 + t - 1$ Circle one: Yes No $\boldsymbol{q(t)} = \_\_\boldsymbol{p_1(t)}\_\_\boldsymbol{p_2(t)}\_\_\boldsymbol{p_3(t)}$

b) $q(t) = t^2$ Circle one: Yes No $\boldsymbol{q(t)} = \_\_\boldsymbol{p_1(t)}\_\_\boldsymbol{p_2(t)}\_\_\boldsymbol{p_3(t)}$

c) $q(t) = 4t^3 - 3t^2 + 5t + 3$ Circle one: Yes No $\boldsymbol{q(t)} = \_\_\boldsymbol{p_1(t)}\_\_\boldsymbol{p_2(t)}\_\_\boldsymbol{p_3(t)}$

4. Let $\boldsymbol{v_1} = \begin{bmatrix} 2 & 1 \\ 1 & 2 \end{bmatrix}$, $\boldsymbol{v_2} = \begin{bmatrix} 1 & 0 \\ 1 & 1 \end{bmatrix}$, and $\boldsymbol{v_3} = \begin{bmatrix} 0 & 1 \\ 2 & 2 \end{bmatrix}$. Determine if each of the following vectors $\boldsymbol{u}$ is a linear combination of $\boldsymbol{v_1}, \boldsymbol{v_2}$, and $\boldsymbol{v_3}$. If it is, then display the linear combination by supplying the coefficients and appropriate operations.

a) $\boldsymbol{u} = \begin{bmatrix} 1 & 0 \\ 0 & 1 \end{bmatrix}$ Circle one: Yes No $\boldsymbol{u} = \_\_\boldsymbol{v_1}\_\_\boldsymbol{v_2}\_\_\boldsymbol{v_3}$

b) $\boldsymbol{u} = \begin{bmatrix} 3 & -1 \\ -2 & -1 \end{bmatrix}$ Circle one: Yes No $\boldsymbol{u} = \_\_\boldsymbol{v_1}\_\_\boldsymbol{v_2}\_\_\boldsymbol{v_3}$

c) $\boldsymbol{u} = \begin{bmatrix} 1 & -2 \\ -3 & -3 \end{bmatrix}$ Circle one: Yes No $\boldsymbol{u} = \_\_\boldsymbol{v_1}\_\_\boldsymbol{v_2}\_\_\boldsymbol{v_3}$

5. Let $\boldsymbol{A} = \begin{bmatrix} 3 & -1 & 0 & 2 \\ 2 & 1 & 2 & 0 \\ -4 & 2 & 3 & 1 \end{bmatrix}$. Express the following linear combinations as a product of a vector and matrix $\boldsymbol{A}$ in an appropriate order by finding a column vector $\boldsymbol{x}$ such that $\boldsymbol{Ax}$ gives the linear combination or row vector $\boldsymbol{y}$ such that $\boldsymbol{yA}$ gives the linear combination. (Note: See Exercise 10 in Lab 3.2.)

a) $-2*\text{col}_1\boldsymbol{A}+3*\text{col}_2\boldsymbol{A}-\text{col}_3\boldsymbol{A}+4*\text{col}_4\boldsymbol{A}$ ______________________

b) $3*\text{col}_1\boldsymbol{A}-\text{col}_2\boldsymbol{A}+5*\text{col}_3\boldsymbol{A}$ ______________________

c) $2*\text{row}_1\boldsymbol{A}-4*\text{row}_2\boldsymbol{A}+3*\text{row}_3\boldsymbol{A}$ ______________________

d) $-5*\text{row}_1\boldsymbol{A}+2*\text{row}_3\boldsymbol{A}$ ______________________

6. Let $\boldsymbol{A}$ be an $m \times n$ matrix and $\boldsymbol{X}$ be a $n \times 1$ matrix. Explain how to write the product $\boldsymbol{A} * \boldsymbol{X}$ as a linear combination of columns of $\boldsymbol{A}$.

____

7. Let $\boldsymbol{A}$ be an $m \times n$ matrix and $\boldsymbol{X}$ be a $1 \times m$ matrix. Explain how to write the product $\boldsymbol{X} * \boldsymbol{A}$ as a linear combination of rows of $\boldsymbol{A}$.

____

8. Let $\mathbf{S} = \{\, \boldsymbol{v}_1, \boldsymbol{v}_2 \,\}$ where $\boldsymbol{v}_1 = \begin{bmatrix} i \\ 1 \end{bmatrix}$ and $\boldsymbol{v}_2 = \begin{bmatrix} -i \\ 1+i \end{bmatrix}$.

a) Write vector $\boldsymbol{x}_1 = \begin{bmatrix} 0 \\ 2+i \end{bmatrix}$ as a linear combination of the elements of $\boldsymbol{S}$.

$\boldsymbol{x}_1 =$ ______ $\boldsymbol{v}_1 +$ ______ $\boldsymbol{v}_2$

b) Write $\boldsymbol{x}_2 = \begin{bmatrix} 1 \\ 1 \end{bmatrix}$ as a linear combination of the elements of $\boldsymbol{S}$.

$\boldsymbol{x}_2 =$ ______ $\boldsymbol{v}_1 +$ ______ $\boldsymbol{v}_2$

Section 6.2

Span

There are two common types of problems related to span. The first is:

Given a set of vectors $\mathbf{S} = \{ \boldsymbol{X_1}, \boldsymbol{X_2}, \ldots, \boldsymbol{X_k} \}$ and a vector $\boldsymbol{X}$, is $\boldsymbol{X}$ in span $\mathbf{S}$?

This is identical to the linear combination problem addressed above because we want to know if $\boldsymbol{X}$ is a linear combination of the members of **S**. As shown in Section 6.1 we can use MATLAB in many cases to solve this problem.

The second type of problem related to span is:

Given a set of vectors $\mathbf{S} = \{ \boldsymbol{X_1}, \boldsymbol{X_2}, \ldots, \boldsymbol{X_k} \}$ in vector space $\mathbf{V}$, does span $\mathbf{S} = \mathbf{V}$?

Here we are asked if every vector in **V** can be written as a linear combination of the vectors in **S**. In this case the corresponding linear system has a right-hand side which contains arbitrary values that correspond to an arbitrary vector in **V**.

For the second type of spanning question there is a special case that arises frequently and can be handled in MATLAB . (The following discussion uses ideas presented in detail later.) The *dimension* of a vector space **V** is the number of vectors in a basis, which is the smallest number of vectors that can span **V**. If we know that **V** has dimension k and set **S** has k vectors, then we can proceed as follows to see if span $\mathbf{S} = \mathbf{V}$. Develop a linear system $\boldsymbol{Ac} = \boldsymbol{b}$ associated with the span question. If the reduced row echelon form of coefficient matrix $\boldsymbol{A}$ has the form

$$\begin{bmatrix} \boldsymbol{I_k} \\ \boldsymbol{O} \end{bmatrix}$$

where $\boldsymbol{O}$ is a submatrix of all zeros, then any vector in **V** is expressible in terms of the members of **S**. (In fact, **S** is a basis for **V**. See Section 6.4.) In MATLAB we can use routines **rowop** or **rref** on the coefficient matrix $\boldsymbol{A}$.

Example 1. Let **V** be a vector space with dim $\mathbf{V} = 3$ and $\mathbf{S} = \left\{ \begin{bmatrix} 1 \\ 1 \\ 0 \\ 2 \end{bmatrix}, \begin{bmatrix} 1 \\ 0 \\ 1 \\ 3 \end{bmatrix}, \begin{bmatrix} -1 \\ 2 \\ 2 \\ 1 \end{bmatrix} \right\}$ be a subset of **V**. Determine if span $\mathbf{S} = \mathbf{V}$.

Since the number of vectors in **S** equals the dimension of **V** we proceed as follows. Let $\boldsymbol{A}$ be the matrix whose columns are the vectors in **S**. Then in MATLAB use command **rref(A)**. We obtain

```
ans =

     1     0     0
     0     1     0
     0     0     1
     0     0     0
```

It follows that set **S** does span **V**.

Example 2. Let **V** be a vector space with dim **V** = 3 and $\mathbf{S} = \left\{ \begin{bmatrix} 1 \\ 1 \\ 0 \\ 2 \end{bmatrix}, \begin{bmatrix} 1 \\ 7 \\ 4 \\ 8 \end{bmatrix}, \begin{bmatrix} -1 \\ 2 \\ 2 \\ 1 \end{bmatrix} \right\}$ be a subset of **V**. Determine if span **S** = **V**.

Since the number of vectors in **S** equals the dimension of **V** we proceed as follows. Let $\boldsymbol{A}$ be the matrix whose columns are the vectors in **S**. Then in MATLAB enter command **rref(A)**. We obtain

```
ans =

    1.0000         0   -1.5000
         0    1.0000    0.5000
         0         0         0
         0         0         0
```

It follows that set **S** does not span **V**.

Frequently we ask if a set S spans R^n. MATLAB can answer this particular case of the second type of spanning question directly. Let $S = \{\boldsymbol{X}_1, \boldsymbol{X}_2, \ldots, \boldsymbol{X}_k\}$ be a set of vectors in R^n. The set will span R^n if every vector $\boldsymbol{X}$ in R^n can be written as a linear combination of the members of S. That is, provided the linear system corresponding to

$$c_1\boldsymbol{X}_1 + c_2\boldsymbol{X}_2 + \ldots + c_k\boldsymbol{X}_k = \boldsymbol{X}$$

is consistent for all possible right hand sides $\boldsymbol{X}$. Recall that a linear system is inconsistent if the rref of its augmented matrix contains a row of the form

$$\left[\begin{array}{cccc|c} 0 & 0 & \ldots & 0 & * \end{array}\right]$$

where $* \neq 0$. By virtue of the row operations $*$ is a linear combination of the entries in $\boldsymbol{X}$. There is no way $*$ can be exactly zero for all possible choices of entries for $\boldsymbol{X}$. Hence if the system is inconsistent not every vector $\boldsymbol{X}$ in R^n can be expressed as a linear combination of the vectors in S. Thus we need only inspect rref of $\boldsymbol{A} = [\boldsymbol{X}_1 \ldots \boldsymbol{X}_k]$ for a zero row to imply S does not span R^n. (Warning: this argument cannot be applied to check spanning sets for proper subspaces.)

Example 3. Determine if

$$S = \left\{ \begin{bmatrix} 1 \\ 1 \\ -1 \end{bmatrix}, \begin{bmatrix} 1 \\ 7 \\ 2 \end{bmatrix}, \begin{bmatrix} 0 \\ 4 \\ 2 \end{bmatrix}, \begin{bmatrix} 2 \\ 8 \\ 1 \end{bmatrix} \right\}$$

spans R^3.

Let $\boldsymbol{A}$ be the matrix whose columns are the vectors in S. Then in MATLAB use **rref(A)** to obtain

```
ans=
       1.0000     0           -0.6667     1.0000
       0          1.0000       0.6667     1.0000
       0          0            0          0
```

Since **rref(A)** contains a row of zeros, it follows by the argument above that S does not span R^3.

Another spanning question involves finding a set that spans the set of solutions of a homogeneous system of equations $\boldsymbol{Ax} = \mathbf{0}$. The strategy in MATLAB is to find the reduced row echelon form of $[\boldsymbol{A} \,|\, \mathbf{0}]$ using the command

rref(A)

(There is no need to include the augmented column since it is all zeros.) Then form the general solution of the system and express it as a linear combination of columns. The columns form a spanning set for the solution set of the system.

Example 4. Determine a spanning set for the set of solutions of the homogeneous system $\boldsymbol{Ax} = \mathbf{0}$ where

$$\boldsymbol{A} = \begin{bmatrix} 1 & 2 & 0 & 1 \\ 2 & 0 & 4 & 2 \\ 3 & -1 & 7 & 1 \end{bmatrix}.$$

In MATLAB find **rref(A)** and write out the general solution as in Lab 4. We obtain

```
ans =

     1     0     2     0
     0     1    -1     0
     0     0     0     1
```

It follows that $x_1 = -2x_3, x_2 = x_3$, and $x_4 = 0$. Let $x_3 = r$, then the general solution is

$$\boldsymbol{x} = \begin{bmatrix} -2r \\ r \\ r \\ 0 \end{bmatrix} = r * \begin{bmatrix} -2 \\ 1 \\ 1 \\ 0 \end{bmatrix}.$$

Hence every vector in the solution space of $\boldsymbol{Ax} = \mathbf{0}$ is a multiple of $\begin{bmatrix} -2 \\ 1 \\ 1 \\ 0 \end{bmatrix}$ and thus the spanning set consists of the single vector $\begin{bmatrix} -2 \\ 1 \\ 1 \\ 0 \end{bmatrix}$. (Later it will be shown that this vector is in fact a basis for the solution space.)

Exercises 6.2

1. Determine if $\boldsymbol{v_1} = \begin{bmatrix} 2 & 1 & 0 \end{bmatrix}$, $\boldsymbol{v_2} = \begin{bmatrix} -1 & 1 & 3 \end{bmatrix}$, $\boldsymbol{v_3} = \begin{bmatrix} 0 & -1 & 6 \end{bmatrix}$ spans the vector space of rows with three real entries which has dimension 3. Record your results below.

2. Determine if $\boldsymbol{v_1} = \begin{bmatrix} 2 \\ 1 \\ 1 \\ 2 \end{bmatrix}$, $\boldsymbol{v_2} = \begin{bmatrix} 1 \\ 1 \\ 0 \\ 1 \end{bmatrix}$, $\boldsymbol{v_3} = \begin{bmatrix} 1 \\ 0 \\ 1 \\ 1 \end{bmatrix}$, $\boldsymbol{v_4} = \begin{bmatrix} 0 \\ 0 \\ 1 \\ 1 \end{bmatrix}$, spans the vector space of columns with four real entries which has dimension 4. Record your results below.

3. Let $\mathbf{S} = \{\boldsymbol{v_1}, \boldsymbol{v_2}, \boldsymbol{v_3}\}$ from Exercise 2. Does $\mathbf{S}$ span a subspace of dimension 3?

Circle one: Yes No Explain your answer in the space below.

4. Let $\mathbf{T} = \{p_1(t),\ p_2(t),\ p_3(t),\ p_4(t)\}$ where $p_1(t) = t+2,\ p_2(t) = t^2 - t,\ p_3(t) = t^3,\ p_4(t) = t^3 - t^2 + 1$. Is span $\mathbf{T}$ = vector space of polynomials of degree 3 or less?

Circle one: Yes No Explain your answer in the space below.

5. Let $\mathbf{S} = \{p_1(t), p_2(t), p_3(t)\}$ from Exercise 4. Does $\mathbf{S}$ span a subspace of dimension 3?

Circle one: Yes No Explain your answer in the space below.

6. Let $\mathbf{T} = \{\boldsymbol{v_1}, \boldsymbol{v_2}, \boldsymbol{v_3}, \boldsymbol{v_4}\}$ where $\boldsymbol{v_1} = \begin{bmatrix} 1 & 0 \\ 1 & 2 \end{bmatrix}$, $\boldsymbol{v_2} = \begin{bmatrix} 1 & 1 \\ 2 & -1 \end{bmatrix}$, $\boldsymbol{v_3} = \begin{bmatrix} 2 & 0 \\ 1 & 4 \end{bmatrix}$,

$\boldsymbol{v_4} = \begin{bmatrix} 1 & 1 \\ 1 & -1 \end{bmatrix}$. Does $\mathbf{T}$ span the vector space of all 2×2 matrices with real entries which has dimension 4?

Circle one: Yes No Explain your answer in the space below.

7. Let $\boldsymbol{v_1} = \begin{bmatrix} 2+3i \\ 4+5i \end{bmatrix}$ and $\boldsymbol{v_2} = \begin{bmatrix} 6-2i \\ 7+i \end{bmatrix}$. Determine whether the set $\boldsymbol{S} = \{\boldsymbol{v_1},\ \boldsymbol{v_2}\}$ spans C^2.

Circle one: YES NO

8. Let $\boldsymbol{v_1} = \begin{bmatrix} 2+3i \\ 4-5i \\ 1+i \end{bmatrix}$, $\boldsymbol{v_2} = \begin{bmatrix} 2-4i \\ 1-i \\ 2i \end{bmatrix}$, and $\boldsymbol{v_3} = \begin{bmatrix} 6+2i \\ 9-11i \\ 2+4i \end{bmatrix}$. Determine whether the set $\boldsymbol{S} = \{\boldsymbol{v_1},\ \boldsymbol{v_2},\ \boldsymbol{v_3}\}$ spans C^3.

Circle one: YES NO

9. Find a spanning set **S** for the set of all solutions of $\boldsymbol{Ax} = \mathbf{0}$ for each of the following matrices $\boldsymbol{A}$. Record the set **S** next to $\boldsymbol{A}$.

a) $\boldsymbol{A} = \begin{bmatrix} 2 & 3 & 1 \\ -1 & 0 & -2 \\ 1 & 2 & 0 \end{bmatrix}$

b) $\boldsymbol{A} = \begin{bmatrix} 1 & 2 & 3 \\ 4 & 5 & 6 \\ 7 & 8 & 0 \end{bmatrix}$

c) $\boldsymbol{A} = \begin{bmatrix} 1 & 2 & 4 & 4 \\ 2 & 1 & -3 & 2 \\ -1 & 3 & 0 & 1 \end{bmatrix}$

d) $\boldsymbol{A} = \begin{bmatrix} 1 & 2 & 4 & 4 & 0 \\ 2 & 1 & -3 & 2 & 1 \\ -1 & 3 & 0 & 1 & 2 \end{bmatrix}$

10. Let $\boldsymbol{v_1} = \begin{bmatrix} 1 \\ 2 \\ 1 \end{bmatrix}$, $\boldsymbol{v_2} = \begin{bmatrix} 2 \\ 1 \\ 3 \end{bmatrix}$. Find a vector $\boldsymbol{v_3}$ so that $\mathbf{T} = \{\boldsymbol{v_1}, \boldsymbol{v_2}, \boldsymbol{v_3}\}$ spans the vector space of all columns with three real entries. Explain your strategy for finding $\boldsymbol{v_3}$ and display it below.

11. Let $\boldsymbol{v_1} = \begin{bmatrix} 3 \\ 1 \\ 0 \end{bmatrix}$. Find two vectors $\boldsymbol{u_1}$ and $\boldsymbol{u_2}$ so that $\mathbf{S} = \{\boldsymbol{v_1}, \boldsymbol{u_1}, \boldsymbol{u_2}\}$ spans the vector space of all columns with three real entries. Explain your strategy for finding $\boldsymbol{u_1}$ and $\boldsymbol{u_2}$ and display them below.

Section 6.3

Linear Independence/Dependence

The independence or dependence of a set of vectors $\mathbf{S} = \{\ \boldsymbol{X_1},\ \boldsymbol{X_2},\ \dots\ ,\ \boldsymbol{X_k}\ \}$ is a linear combination question. A set **S** is *linearly independent* if the only time the linear combination

$$c_1\boldsymbol{X_1} + c_2\boldsymbol{X_2} + \cdots + c_k\boldsymbol{X_k}$$

gives the zero vector is when $c_1 = c_2 = \cdots = c_k = 0$. If we can produce the zero vector with any one of the coefficients $c_j \neq 0$, then **S** is *linearly dcpendent.* From the expression

$$c_1\boldsymbol{X_1} + c_2\boldsymbol{X_2} + \cdots + c_k\boldsymbol{X_k}=\mathbf{0}$$

we derive a homogeneous linear system $\boldsymbol{Ac} = \mathbf{0}$ as we did in Section 6.2 for linear combination problems. Then we have the following result:

S is linearly independent if and only if $\boldsymbol{Ac} = \mathbf{0}$ has only the trivial solution.

Otherwise, **S** is linear dependent. Once we have the homogeneous system $\boldsymbol{Ac} = \mathbf{0}$ we can use MATLAB routines **rowop** or **rref** to analyze whether or not the system has a nontrivial solution.

Example 1. Let $\boldsymbol{X_1}= (1,2,1,-1)$, $\boldsymbol{X_2}= (1,0,2,-3)$, and $\boldsymbol{X_3}= (1,1,0,-2)$. Determine if the vectors in $\mathbf{S} =\{\boldsymbol{X_1},\ \boldsymbol{X_2},\ \boldsymbol{X_3}\}$ are linearly independent or linearly dependent.

Form the expression

$$\sum_{j=1}^{3} c_j\boldsymbol{X_j} = c_1\boldsymbol{X_1} + c_2\boldsymbol{X_2} + c_3\boldsymbol{X_3} = \mathbf{0}$$

and find the corresponding linear system to solve for the c_i. We have

$$c_1(1,2,1,-1) + c_2(1,0,2,-3) + c_3(1,1,0,-2) = (0,0,0,0)$$

Performing the scalar multiplications and adding corresponding entries gives

$$(c_1 + c_2 + c_3,\ \ 2c_1 + c_3,\ \ c_1 + 2c_2,\ \ -c_1 - 3c_2 - 2c_3) = (0,0,0,0)$$

Since the two vectors are equal, we equate corresponding entries to obtain the system of equations

$$\begin{aligned} c_1 + c_2 + c_3 &= 0 \\ 2c_1 \qquad + c_3 &= 0 \\ c_1 + 2c_2 \qquad &= 0 \\ -c_1 - 3c_2 - 2c_3 &= 0 \end{aligned}$$

The matrix form for this linear system is $\boldsymbol{Ac} = \mathbf{0}$ where

$$A = \begin{bmatrix} 1 & 1 & 1 \\ 2 & 0 & 1 \\ 1 & 2 & 0 \\ -1 & -3 & -2 \end{bmatrix} \text{ and } \mathbf{c} = \begin{bmatrix} c_1 \\ c_2 \\ c_3 \end{bmatrix}$$

Enter A into MATLAB and then use command

rref(A)

to give

```
ans =

     1     0     0
     0     1     0
     0     0     1
     0     0     0
```

Recall that this display represents the reduced row echelon form of the coefficient matrix of a homogeneous system. It follows that $c_1 = 0,\ c_2 = 0,\ c_3 = 0$. Hence the set **S** is linearly independent. (Compare this problem with Example 1 in Section 6.1.)

A special case arises if we have k vectors in a set **S** in a vector space **V** whose dimension is k. Let the linear system associated with the linear combination problem be $A\mathbf{c} = \mathbf{0}$. It can be shown that

S is linearly independent if and only if the reduced row echelon form of A is $\begin{bmatrix} I_k \\ O \end{bmatrix}$

where O is a submatrix of all zeros. In fact we can extend this further to say **S** is a basis for **V**. (See Section 6.4.) In MATLAB we can use **rowop** or **rref** on A to aid in the analysis of such a situation.

The command **lisub** can be very efficient when dealing with problems on linear independence/dependence. Type **help lisub** or see Section 12.3, for more details.

Exercises 6.3

1. Determine if the following sets are linearly independent or linearly dependent. Record your findings in the space provided.

a) S $=\{\boldsymbol{v_1} = [4 \quad 2 \quad 1]\ ,\ \boldsymbol{v_2} = [-2 \quad 3 \quad 1]\ ,\ \boldsymbol{v_3} = [2 \quad -11 \quad -4]\ \}$

---

b) S $=\{\boldsymbol{v_1} = [3 \quad 1 \quad 2]\ ,\ \boldsymbol{v_2} = [-1 \quad 1 \quad 3]\ ,\ \boldsymbol{v_3} = [7 \quad 1 \quad 1]\ \}$

---

c) $\mathbf{S} = \left\{ \boldsymbol{v_1} = \begin{bmatrix} 1 \\ 2 \\ 1 \\ -2 \end{bmatrix}, \boldsymbol{v_2} = \begin{bmatrix} 2 \\ 1 \\ -3 \\ -1 \end{bmatrix}, \boldsymbol{v_3} = \begin{bmatrix} 1 \\ 2 \\ 6 \\ -5 \end{bmatrix} \right\}$

---

d) $\mathbf{S} = \left\{ \boldsymbol{v_1} = \begin{bmatrix} 1 \\ -1 \\ 2 \\ 4 \end{bmatrix}, \boldsymbol{v_2} = \begin{bmatrix} 0 \\ 2 \\ 1 \\ 1 \end{bmatrix}, \boldsymbol{v_3} = \begin{bmatrix} 3 \\ 1 \\ 0 \\ 2 \end{bmatrix} \right\}$

---

e) S $=\{p_1(t) = t^2 + 2t + 1,\ \ p_2(t) = t + 2,\ \ p_3(t) = 3t^2 + 4t - 1\}$

---

f) $\mathbf{S} = \left\{ \boldsymbol{v_1} = \begin{bmatrix} 2 & 1 \\ 1 & 2 \end{bmatrix}, \boldsymbol{v_2} = \begin{bmatrix} 1 & 0 \\ 1 & 1 \end{bmatrix}, \boldsymbol{v_3} = \begin{bmatrix} 0 & -1 \\ -1 & 0 \end{bmatrix} \right\}$

---

2. Let $\boldsymbol{v_1} = \begin{bmatrix} 2+3i \\ 4+5i \end{bmatrix}$, $\boldsymbol{v_2} = \begin{bmatrix} 6-2i \\ 7+i \end{bmatrix}$, and $\boldsymbol{v_3} = \begin{bmatrix} 6+12i \\ 7+17i \end{bmatrix}$. Is the set $\boldsymbol{S} = \{\boldsymbol{v_1},\ \boldsymbol{v_2},\ \boldsymbol{v_3}\}$ linearly independent?

Circle one: YES NO

3. Let $\boldsymbol{v_1} = \begin{bmatrix} 2+3i \\ 4-5i \\ 1+i \end{bmatrix}$ and $\boldsymbol{v_2} = \begin{bmatrix} 2-4i \\ 1-i \\ 2i \end{bmatrix}$. Is the set $\boldsymbol{S} = \{\boldsymbol{v_1},\ \boldsymbol{v_2}\ \}$ linearly independent?

Circle one: YES NO

Section 6.4

<u>Basis</u>

A set **S** = { $\boldsymbol{X_1}$, $\boldsymbol{X_2}$, ... , $\boldsymbol{X_k}$ } is a basis for a vector space **V** provided span **S** = **V** <u>and</u> **S** is linearly independent. Thus we use the strategies discussed previously for checking span problems and those for linear independence problems. However, there are a number of special cases which arise often enough that we single them out.

- The *row space* of a matrix is the span of its rows. For an m × n matrix $\boldsymbol{B}$, the nonzero rows of **rref(B)** form a basis for the row space of $\boldsymbol{B}$.

- The *column space* of a matrix is the span of its columns. For an m × n matrix $\boldsymbol{B}$, the nonzero columns of (**rref(B′)**)′ form a basis for the column space of $\boldsymbol{B}$.

- The rows (columns) of a square matrix $\boldsymbol{A}$ form a basis for the row (column) space of $\boldsymbol{A}$ if and only if **rref(A)** = $\boldsymbol{I}$.

- A basis for the solution space of $\boldsymbol{Ax} = \mathbf{0}$, often called the *null space* of matrix $\boldsymbol{A}$, can be found by using **rref(A)** or **homsoln(A)** to write out the general solution. (See Section 12.4 for a description of the routine **homsoln**.)

- Let dim **V** = k and **S** = { $\boldsymbol{X_1}$, $\boldsymbol{X_2}$, ... , $\boldsymbol{X_k}$ } be a subset of **V**.

  i) If **S** is linearly independent, then **S** is a basis for **V**. (See the discussion of Independence/Dependence.)

  ii) If span **S** = **V**, then **S** is a basis for **V**. (See the discussion of Span.)

- If dim **V** = k, then any set with <u>fewer</u> than k vectors cannot be a basis for **V**.

- If dim **V** = k, then any set with <u>more</u> than k vectors cannot be a basis for **V**.

<u>Example 1.</u> Let $\boldsymbol{A}=\begin{bmatrix} 1 & 1 & 2 \\ 2 & 0 & 1 \\ 1 & 1 & 0 \end{bmatrix}$. Do the columns of $\boldsymbol{A}$ form a basis for the vector space of all columns with three real entries, which is often denoted by R^3?

Since dim R^3 = 3 and $\boldsymbol{A}$ has 3 columns, we need only check that the columns of $\boldsymbol{A}$ are linearly independent. In MATLAB command **rref(A)** displays

```
ans =

     1     0     0
     0     1     0
     0     0     1
```

It follows that the columns of $\boldsymbol{A}$ are linearly independent and hence are a basis for R^3. In addition we can also conclude that the rows of $\boldsymbol{A}$ are a basis for the row space.

Example 2. Find a basis for the row space and a basis for the column space of $\boldsymbol{A} = \begin{bmatrix} 1 & 2 & 1 & 3 & 1 \\ 2 & 1 & 5 & -3 & 1 \\ 1 & 2 & 1 & 3 & 1 \\ 3 & 0 & 9 & -9 & 1 \\ 1 & 1 & 2 & 0 & 1 \end{bmatrix}$.

Enter the matrix into MATLAB and use the following commands. For the row space **rref(A)** displays

```
ans =

     1     0     3    -3     0
     0     1    -1     3     0
     0     0     0     0     1
     0     0     0     0     0
     0     0     0     0     0
```

which implies that the 3 nonzero rows are a basis. Similarly, command **rref(A′)′** displays

```
ans =

     1     0     0     0     0
     0     1     0     0     0
     1     0     0     0     0
    -1     2     0     0     0
     0     0     1     0     0
```

which implies that the 3 nonzero columns are a basis for the column space.

Lab 12 investigates the kernel and range of a linear transformation in terms of the null space and column space of a matrix. You may want to apply two commands that are introduced there - **lisub** and **homsoln** - to the concepts presented in the present lab. Use **help** for more information on these commands.

Exercises 6.4

1. Let $\mathbf{V} = \mathrm{R}^3$. Determine whether the following sets are a basis for $\mathbf{V}$. It may be possible to decide without any computations. Record your response next to each set.

a) $\mathbf{S} = \left\{ \begin{bmatrix} 1 \\ 2 \\ 1 \end{bmatrix}, \begin{bmatrix} 2 \\ 1 \\ 1 \end{bmatrix}, \begin{bmatrix} 0 \\ 3 \\ 1 \end{bmatrix} \right\}$

b) $\mathbf{S} = \left\{ \begin{bmatrix} 1 \\ 1 \\ 0 \end{bmatrix}, \begin{bmatrix} 1 \\ 0 \\ 1 \end{bmatrix}, \begin{bmatrix} 0 \\ 1 \\ 1 \end{bmatrix} \right\}$

c) $\mathbf{S} = \left\{ \begin{bmatrix} 3 \\ 1 \\ 3 \end{bmatrix}, \begin{bmatrix} 3 \\ 1 \\ 2 \end{bmatrix} \right\}$

d) $\mathbf{S} = \left\{ \begin{bmatrix} 3 \\ 1 \\ 3 \end{bmatrix}, \begin{bmatrix} 3 \\ 1 \\ 2 \end{bmatrix}, \begin{bmatrix} 3 \\ 1 \\ 1 \end{bmatrix}, \begin{bmatrix} 3 \\ 1 \\ 0 \end{bmatrix} \right\}$

2. Let $\boldsymbol{A} = \begin{bmatrix} 2 & 1 & 3 & -1 & 1 \\ 3 & 1 & 0 & 1 & 0 \\ 1 & 2 & 1 & 1 & 2 \end{bmatrix}$.

a) Find a basis for the row space of $\boldsymbol{A}$.

b) Find a basis for the column space of $\boldsymbol{A}$.

c) Find a basis for the solution space of $\boldsymbol{Ax} = \mathbf{0}$.

3. Let $\mathbf{S} = \{\boldsymbol{v}_1, \boldsymbol{v}_2\}$ where $\boldsymbol{v}_1 = \begin{bmatrix} i \\ 1 \end{bmatrix}$ and $\boldsymbol{v}_2 = \begin{bmatrix} -i \\ 1+i \end{bmatrix}$. Determine if $\mathbf{S}$ is a basis for C^2, the set of all column vectors with two complex entries. Explain your answer in the space below.

Section 6.5

Symbolic Operations

The routine **symrowop** can be used to solve many types of problems (involving concepts presented in this lab) which cannot be solved using routines in regular MATLAB . We present an example to illustrate one such problem. Recall that **symrowop** requires the Symbolic Math Toolbox.

Example 1. Find all values of a for which the set of vectors $\{\boldsymbol{v1}, \boldsymbol{v2}, \boldsymbol{v3}\}$ is linearly independent:

$$\boldsymbol{v1} = \begin{bmatrix} a \\ a+2 \\ a+1 \end{bmatrix}, \quad \boldsymbol{v2} = \begin{bmatrix} a+1 \\ a \\ a+2 \end{bmatrix}, \quad \boldsymbol{v3} = \begin{bmatrix} a+2 \\ a+1 \\ a \end{bmatrix}$$

Each vector $\boldsymbol{vi}$ lies in $\boldsymbol{R}^3$ so the set $\{\boldsymbol{v1}, \boldsymbol{v2}, \boldsymbol{v3}\}$ is linearly independent whenever the matrix $\boldsymbol{A}$, whose columns are the vectors $\boldsymbol{vi}$, reduces to the identity matrix $\boldsymbol{I}$. We show under what conditions $\boldsymbol{A}$ reduces to $\boldsymbol{I}$.

Type **symrowop** and enter

$$\mathbf{sym}('[\ a,\ \ a+1,\ \ a+2;\ \ a+2,\ \ a,\ \ a+1;\ \ a+1,\ \ a+2,\ \ a\ ]')$$

It is convenient to begin by producing $A(1,1) = 1$, so we set about multiplying **Row(1)** by $k = 1/a$. The first step is to type $1/a$ in the box beside **k** = under the row operation **k*Row(i)**. However, as soon as you press the Tab key or click the mouse in the box beside **i** = the Comment Window displays a message with two lines:

Division; denominator assumed not ZERO Acknowledge restriction; click on CONTINUE

The reason for this warning is to alert you to the fact that MATLAB performs such an operation without regard to this restriction. Here it is important for you to note that multiplying by $1/a$ assumes that $a \neq 0$. To continue with **symrowop** click the CONTINUE button, which is displayed in a red border to emphasize that a warning has been issued. Next you must click in the appropriate box again before entering $i = 1$. This displays

$$\begin{bmatrix} 1 & 1/a*(a+1) & 1/a*(a+2) \\ a+2 & a & a+1 \\ a+1 & a+2 & a \end{bmatrix}$$

(We have included matrix brackets here, but they do not appear in the screen displays.) A comment on MATLAB 's order of operations is appropriate here. Look at the value of $A(1,2)$. It means

$$\frac{1}{a}(a+1)$$

since MATLAB uses the standard order of numeric operations, sometimes remembered by **PEMDAS**[1] where the order of precedence is PE(MD)(AS). This means that multiplication and division are on the same level, with addition and subtraction carried out after all multiplications and divisions have been performed. For operations on the same level, the order of execution is taken from left to right.

The next step is to produce $A(2,1) = 0$ by the operation **k*Row(i) + Row(j)**, with $k = -(a+2)$, $i = 1$, and $j = 2$. Proceed similarly to make $A(3,1) = 0$. After these operations the matrix displayed is

$$\begin{bmatrix} 1 & 1/a*(a+1) & 1/a*(a+2) \\ 0 & -(3*a+2)/a & -(3*a+4)/a \\ 0 & -1/a & -(3*a+2)/a \end{bmatrix}$$

At this point it is convenient, although not necessary, to switch **Row(2)** and **Row(3)**. Then multiply **Row(2)** by $k = -a$. Finally, produce $A(3,2) = 0$ using $k = (3*a+2)/a$, $i = 2$, and $j = 3$, to obtain the row echelon form

$$\begin{bmatrix} 1 & 1/a*(a+1) & 1/a*(a+2) \\ 0 & 1 & 3*a+2 \\ 0 & 0 & 9*a+9 \end{bmatrix}$$

This completes MATLAB 's role. Now the user must interpret this form. Clearly if $a = -1$ then $A(3,3) = 0$, otherwise $A(3,3) \neq 0$. Therefore the matrix $\boldsymbol{A}$ reduces to the identity matrix $\boldsymbol{I}$ whenever $a \neq -1$, so the set $\{\boldsymbol{v1}, \boldsymbol{v2}, \boldsymbol{v3}\}$ is linearly independent when $a \neq -1$.

What about the assumption that **symrowop** forced us to acknowledge, namely, $a \neq 0$? To check this case, as well as to check the answer to Example 1, quit **symrowop**. Enter $a = 0$, then define the matrix

[1] A mnemonic for this is 'Please Excuse My Dear Aunt Sally', and the letters respectively denote Parentheses, Exponentiation, Multiplication & Division, Addition & Subtraction.

M= [a $a+1$ $a+2$; $a+2$ a $a+1$; $a+1$ $a+2$ a]

The command **rref(M)** displays the identity matrix, which indicates that the set of vectors $\{\boldsymbol{v1}, \boldsymbol{v2}, \boldsymbol{v3}\}$ is linearly independent when $a = 0$. Therefore the conclusion is that $\{\boldsymbol{v1}, \boldsymbol{v2}, \boldsymbol{v3}\}$ is linearly independent for all $a \neq -1$.

To verify this conclusion, define $a = -1$, then enter matrix **M** again. (Use the MATLAB Command Stack; up arrow.) The command **rref(M)** displays

```
1   0   -1
0   1   -1
0   0    0
```

This means that the set $\{\boldsymbol{v1}, \boldsymbol{v2}, \boldsymbol{v3}\}$ is linearly dependent with $\boldsymbol{v1} + \boldsymbol{v2} + \boldsymbol{v3} = \mathbf{0}$.

Exercises 6.5

1. For what value of t will $\boldsymbol{b}$ be a linear combination of $\boldsymbol{x1}$ and $\boldsymbol{x2}$, where

$$\boldsymbol{x1} = \begin{bmatrix} 1 \\ -2 \\ 4 \end{bmatrix}, \quad \boldsymbol{x2} = \begin{bmatrix} -2 \\ 5 \\ 3 \end{bmatrix}, \quad \boldsymbol{b} = \begin{bmatrix} -3 \\ 8 \\ t \end{bmatrix}$$

$t =$ ____________________

2. For what value of t will $\boldsymbol{b}$ be a linear combination of $\boldsymbol{x1}$ and $\boldsymbol{x2}$, where

$$\boldsymbol{x1} = \begin{bmatrix} 1 \\ -2 \\ 0 \end{bmatrix}, \quad \boldsymbol{x2} = \begin{bmatrix} -2 \\ 7 \\ 1 \end{bmatrix}, \quad \boldsymbol{b} = \begin{bmatrix} t \\ -5 \\ -3 \end{bmatrix}$$

$t =$ ____________________

3. Find all values of r and s such that the vector $\begin{bmatrix} r \\ s \\ s-r \end{bmatrix}$ will be in Span$\{\boldsymbol{x1},\boldsymbol{x2}\}$, where

$$\boldsymbol{x1} = \begin{bmatrix} 1 \\ 2 \\ 1 \end{bmatrix}, \quad \boldsymbol{x2} = \begin{bmatrix} -1 \\ 2 \\ 3 \end{bmatrix}$$

$r =$ ______________ $s =$ ______________

4. Find all values of r and s such that the vector $\begin{bmatrix} r \\ s \\ s+r \end{bmatrix}$ will be in Span$\{\boldsymbol{x1}, \boldsymbol{x2}\}$, where

$$\boldsymbol{x1} = \begin{bmatrix} 1 \\ 2 \\ 1 \end{bmatrix}, \quad \boldsymbol{x2} = \begin{bmatrix} -1 \\ 2 \\ 3 \end{bmatrix}$$

$r =$ ______________ $s =$ ______________

5. For what values of t will $\boldsymbol{y}$ be in Span$\{\ \boldsymbol{v1},\ \boldsymbol{v2},\ \boldsymbol{v3}\ \}$, if

$$\boldsymbol{v1} = \begin{bmatrix} 1 \\ 0 \\ 2 \end{bmatrix}, \quad \boldsymbol{v2} = \begin{bmatrix} 4 \\ 1 \\ -7 \end{bmatrix}, \quad \boldsymbol{v3} = \begin{bmatrix} -1 \\ -2 \\ 0 \end{bmatrix}, \quad \boldsymbol{y} = \begin{bmatrix} 4 \\ -1 \\ t \end{bmatrix}$$

$t =$ ______________

6. For what values of t is the set $\{\ \boldsymbol{v1},\ \boldsymbol{v2},\ \boldsymbol{v3}\ \}$ linearly independent?

$$\boldsymbol{v1} = \begin{bmatrix} 1 \\ 3 \\ 3 \end{bmatrix}, \quad \boldsymbol{v2} = \begin{bmatrix} -2 \\ -4 \\ 1 \end{bmatrix}, \quad \boldsymbol{v3} = \begin{bmatrix} -1 \\ 1 \\ t \end{bmatrix}$$

$t =$ ______________

<< NOTES; COMMENTS; IDEAS >>

LAB 7

Coordinates and Change of Basis

Topics: routine **lincombo**; coordinates of a vector relative to a basis; change of coordinates from one basis to another using a transition matrix.

Introduction

The notion of a vector written as a linear combination of a given set of vectors is one of the most fundamental concepts in a linear algebra course. In Lab 2 and Lab 4 we saw how the routines **rowop**, **symrowop**, and **rref** made essential use of linear combinations of rows to solve a linear system of equations. Exercise 10 in Section 3.2 investigated the form of the matrix produce $\boldsymbol{A} * \boldsymbol{x}$ as a linear combination of the columns of $\boldsymbol{A}$. Section 6.1 was devoted entirely to linear combinations of various kinds of vectors (rows, columns, matrices, and polynomials), while the rest of Lab 6 applied this concept to the notions of span, linear independence, basis, and dimension.

The aim of this lab is to extend linear combinations to the notion of coordinates. Section 7.1 provides an intuitive introduction that uses the routine **lincombo** to reinforce the geometric underpinning of coordinates in R^2. This section could have appeared before Lab 6 without loss of understanding, except that here we refer to a basis. There are no exercises following the section; all the exercises are contained within the routine **lincombo**.

Section 7.2 introduces formally the coordinates of a vector relative to a given basis. Section 7.3 investigates the question, 'How are the coordinates relative to different bases related?' Neither section requires Section 7.1.

Section 7.1

The Linear Combination Game

In this section we introduce the routine **lincombo** to aid in visualizing how a given vector is a linear combination of vectors. **Lincombo** is a graphics game to express one vector as a linear combination of two other vectors. You may type **help lincombo** for directions on using this routine. Type **lincombo** to initiate the routine. A randomly selected trio of vectors is displayed. The initial screen presents two basis vectors, $\boldsymbol{u}$ (in red) and $\boldsymbol{v}$ (in blue). It also shows a third vector, which we shall denote by $\boldsymbol{t}$, in a contrasting color. The objective is to find scalars $c1$ and $c2$ such that

$$c1 * \boldsymbol{u} + c2 * \boldsymbol{v} = \boldsymbol{t}$$

Geometrically the object is to 'size' a parallelogram in such a way that the given vector becomes the diagonal of that parallelogram.

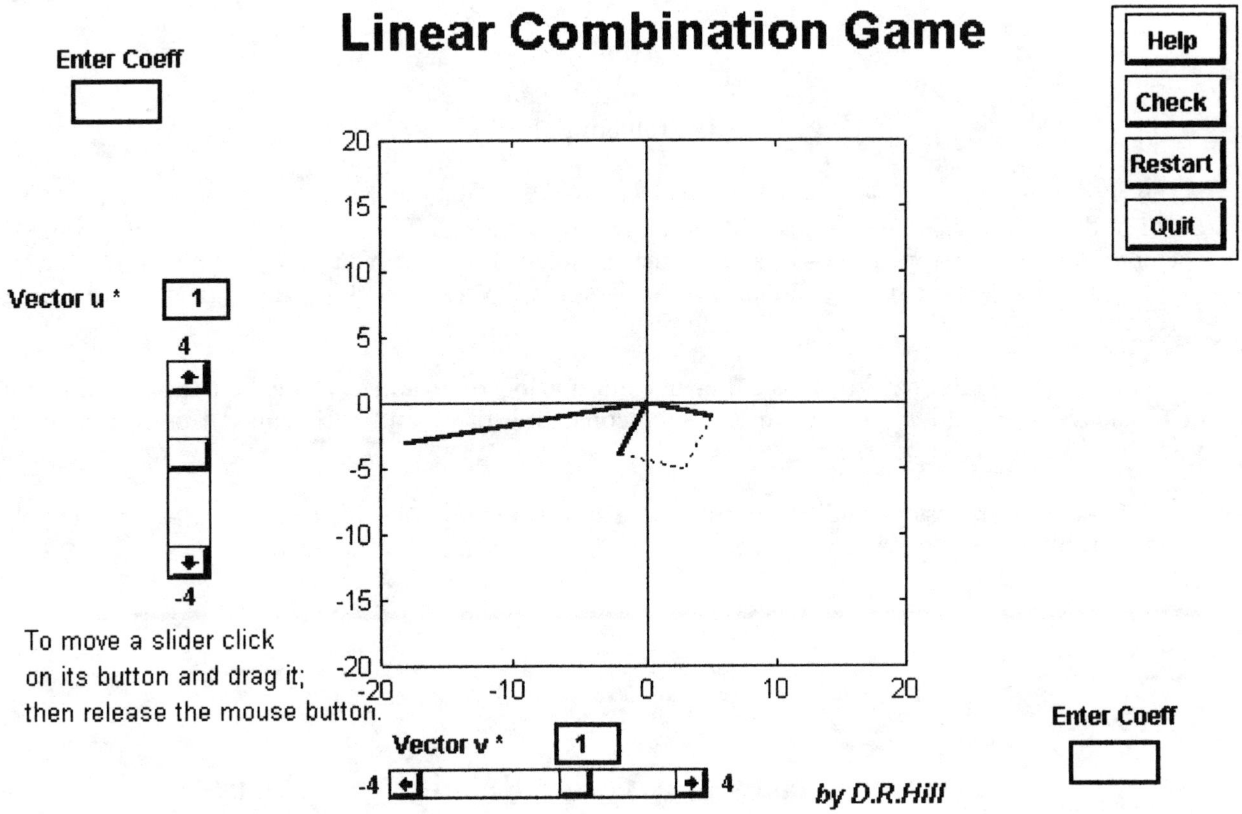

Figure 1.

Begin by varying the values of $c1$ until the length of $c1*\boldsymbol{u}$ seems to form one side of the parallelogram. This is done in either of two ways: by entering coefficients directly or by using the slider near 'Vector $u*$'. Initially it is better to use the slider. (See Figure 1.) Directions for using them appear on the screen. Adjust the slider for the coefficient of $\boldsymbol{u}$. The corresponding value of $c1$ is shown inside the box 'Vector $u*$' located above the slider.

Once the value of $c1$ is approximated, repeat this process for $c2$ using the other slider. You may have to refine your estimates for $c1$ and $c2$ several times until the vector $\boldsymbol{t}$ appears to be the diagonal of the parallelogram. We have restricted the answers for $c1$ and $c2$ to be in tenths between -4 and 4. This enables you to concentrate on the concept of a linear combination without unnecessary distraction. Use this restriction to determine an initial estimate for $c1$ and $c2$. Enter these values in the corresponding boxes under 'Enter Coeff'. Press ENTER to redraw the parallelogram.

If the vector $\boldsymbol{t}$ does not seem to be an exact diagonal of the parallelogram, continue to refine your estimate. When it appears to be exact, press the Check button in the upper-right corner of the screen. The routine responds with one of three messages:

(1) Keep trying.
(2) Close; try again.
(3) Problem Solved.

If the response is (1) your estimate is not very close, (2) you should continue to enter new values for c1 and/or c2; you are quite close, (3) you have the solution. The message 'Close; try again.' is accompanied by the appearance of a new button called Solution. You may click on the Solution button and the correct linear combination will be displayed for a short time. You should then enter the correct coefficients to verify the result.

The scalars which express $\boldsymbol{t}$ as a linear combination of $\boldsymbol{u}$ and $\boldsymbol{v}$ are called the coordinates of $\boldsymbol{t}$ relative to $\boldsymbol{u}$ and $\boldsymbol{v}$. The column vector consisting of $c1$ and $c2$ is called the **coordinate vector** of $\boldsymbol{t}$ relative to $\boldsymbol{u}$ and $\boldsymbol{v}$.

Repeat this process of finding coordinates using **lincombo** by pressing the Restart button. A new problem will be generated. To stop the routine click the Quit button.

Section 7.2

The Coordinates of a Vector Relative to a Basis

The usual coordinate system for R^2 involves basis vectors

$$\boldsymbol{e_1} = \begin{bmatrix} 1 \\ 0 \end{bmatrix} \quad \text{and} \quad \boldsymbol{e_2} = \begin{bmatrix} 0 \\ 1 \end{bmatrix}$$

Any vector $\boldsymbol{v} = \begin{bmatrix} a \\ b \end{bmatrix} = a\boldsymbol{e_1} + b\boldsymbol{e_2}$. We say the coordinates of vector $\boldsymbol{v}$ are its components. This agrees with the idea that $\boldsymbol{v}$ in R^2 can be considered a vector or a point which has coordinates (a,b). It is so easy and natural in this case because we are using the natural basis for R^2.

We want a more general approach that permits us to use other basis vectors in place of the natural basis. To use other bases **we define the coordinates of a vector relative to a basis S to be the scalars used to write the vector as a linear combination of the basis vectors.** We really consider an **ordered basis**; that is, if we switch the order of the basis vectors we get a new basis. Hence **we get a unique set of coordinates relative to an ordered basis.** This correspondence between vectors and coordinates allows us to model an abstract vector space using R^n.

Let $\boldsymbol{V}$ be a vector space with (ordered) basis $\boldsymbol{S} = \{\boldsymbol{v_1}, \boldsymbol{v_2}, \ldots, \boldsymbol{v_n}\}$. For any vector $\boldsymbol{w}$ in $\boldsymbol{V}$ there exist unique scalars $k_1, k_2, \ldots, k_n$ so that

$$\boldsymbol{w} = k_1\boldsymbol{v_1} + k_2\boldsymbol{v_2} + \cdots + k_n\boldsymbol{v_n} \tag{7.1}$$

The scalars $k_1, k_2, \ldots, k_n$ are the coordinates of $\boldsymbol{w}$ relative to the $\boldsymbol{S}$ basis, denoted by $[\boldsymbol{w}]_S$ and written

$$[\boldsymbol{w}]_S = \begin{bmatrix} k_1 \\ k_2 \\ \vdots \\ k_n \end{bmatrix}.$$

To determine the vector $[\boldsymbol{w}]_S$ we must solve Equation (7.1) for $k_1, k_2, \ldots, k_n$. Thus we can say:

> Finding coordinates relative to a basis **S** is a linear combination problem.

Example 1. $\mathbf{S} = \{\boldsymbol{v_1}, \boldsymbol{v_2}\} = \left\{ \begin{bmatrix} 1 \\ 1 \end{bmatrix}, \begin{bmatrix} -1 \\ 2 \end{bmatrix} \right\}$ is a basis for R^2. (Tell how to prove this.) Find the coordinates of vector $\boldsymbol{v} = \begin{bmatrix} -1 \\ 8 \end{bmatrix}$ relative to basis **S**. We seek scalars k_1 and k_2 so that linear combination

$$k_1\boldsymbol{v_1} + k_2\boldsymbol{v_2} = k_1 \begin{bmatrix} 1 \\ 1 \end{bmatrix} + k_2 \begin{bmatrix} -1 \\ 2 \end{bmatrix} = \begin{bmatrix} -1 \\ 8 \end{bmatrix} = \boldsymbol{v}$$

This leads us to solve the linear system whose augmented matrix is

$$\left[\begin{array}{cc|c} 1 & -1 & -1 \\ 1 & 2 & 8 \end{array}\right]$$

In MATLAB if we enter the coefficient matrix $\boldsymbol{A} = \begin{bmatrix} 1 & -1 \\ 1 & 2 \end{bmatrix}$ and right-hand side $\boldsymbol{b} = \begin{bmatrix} -1 \\ 8 \end{bmatrix}$, the solution of the system is given by the MATLAB command

x = A\ b

which gives

```
x =

    2
    3
```

So $k_1 = 2$ and $k_2 = 3$. We have the coordinates of $\boldsymbol{v}$ relative to the $\boldsymbol{S}$ basis as

$$[\boldsymbol{v}]_S = \begin{bmatrix} 2 \\ 3 \end{bmatrix}.$$

The notion of a coordinate of a vector relative to a basis **S** can be generalized to any vector space V. (From now on when we say basis we will mean an ordered basis.) **The computation of coordinates is always a linear combination problem.** The linear system derived may have to be obtained in a different way depending on the vectors of V, but once it is obtained we proceed as in the previous example.

Exercises 7.2

1. In R^3, $\mathbf{S} = \{\boldsymbol{v_1},\boldsymbol{v_2},\boldsymbol{v_3}\} = \left\{ \begin{bmatrix} 1 \\ 1 \\ 2 \end{bmatrix}, \begin{bmatrix} 2 \\ 1 \\ 1 \end{bmatrix}, \begin{bmatrix} 1 \\ 2 \\ 1 \end{bmatrix} \right\}$ is a basis. Find the coordinate vectors of each of the following:

For $\boldsymbol{v} = \begin{bmatrix} 1 \\ 1 \\ 1 \end{bmatrix}$, $[\boldsymbol{v}]_S =$ ____________

For $\boldsymbol{w} = \begin{bmatrix} 1 \\ 0 \\ 1 \end{bmatrix}$, $[\boldsymbol{w}]_S =$ ____________

For $\boldsymbol{u} = \boldsymbol{v} + \boldsymbol{w}$, $[\boldsymbol{u}]_S =$ ____________

Is it true that $[\boldsymbol{v} + \boldsymbol{w}]_S = [\boldsymbol{v}]_S + [\boldsymbol{w}]_S$? YES NO (Circle one.)

For $\boldsymbol{y} = 8\boldsymbol{v}$, $[\boldsymbol{y}]_S =$ ____________

Is it true that $[8\boldsymbol{v}]_S = 8\,[\boldsymbol{v}]_S$? YES NO (Circle one.)

2. In R^4, $\mathbf{T} = \{\boldsymbol{v_1},\boldsymbol{v_2},\boldsymbol{v_3}\ \boldsymbol{v_4}\} = \left\{ \begin{bmatrix} 1 \\ 1 \\ 0 \\ 1 \end{bmatrix}, \begin{bmatrix} 1 \\ 2 \\ 1 \\ 0 \end{bmatrix} \begin{bmatrix} 0 \\ 1 \\ 2 \\ 1 \end{bmatrix} \begin{bmatrix} -1 \\ 0 \\ 0 \\ 1 \end{bmatrix} \right\}$ is a basis. Find the coordinate vectors of each of the following:

For $\boldsymbol{v} = \begin{bmatrix} 2 \\ 1 \\ 1 \\ 2 \end{bmatrix}$, $[\boldsymbol{v}]_T =$ ____________

For $\boldsymbol{w} = \begin{bmatrix} 1 \\ 2 \\ 3 \\ 4 \end{bmatrix}$, $[\boldsymbol{w}]_T =$ ____________

For $\boldsymbol{u} = \boldsymbol{v} + \boldsymbol{w}$, $[\boldsymbol{u}]_T =$ ____________

Is it true that $[\boldsymbol{v} + \boldsymbol{w}]_T = [\boldsymbol{v}]_T + [\boldsymbol{w}]_T$? YES NO (Circle one.)

For $\boldsymbol{y} = -4\boldsymbol{v}$, $[\boldsymbol{y}]_T =$ ____________

Is it true that $[-4\boldsymbol{v}]_T = -4\,[\boldsymbol{v}]_T$? YES NO (Circle one.)

3. Let $\mathbf{S} = \{-x+1, x+1, x^2+x+1\}$ be a basis for the vector space of polynomials of degree 2 or less.

For polynomial $p(x) = -3x^2 + x + 2$, the coordinate vector is

$[\boldsymbol{p(x)}]_S =$ ____________

For polynomial $q(x) = x$, the coordinate vector is

$[\boldsymbol{q(x)}]_S =$ ____________

4. In C^2, $\mathbf{S} = \left\{ \begin{bmatrix} i \\ 1 \end{bmatrix}, \begin{bmatrix} -i \\ 1+i \end{bmatrix} \right\}$ is a basis. Find the coordinate vectors for each of the following.

a) $\boldsymbol{v}= \begin{bmatrix} 0 \\ 2+i \end{bmatrix}$ $[\boldsymbol{v}]_S =$ ________ **b)** $\boldsymbol{v}= \begin{bmatrix} 1 \\ 1 \end{bmatrix}$ $[\boldsymbol{v}]_S =$ ________

5. In C^3, $\mathbf{S} = \left\{ \begin{bmatrix} 1+i \\ 0 \\ 0 \end{bmatrix}, \begin{bmatrix} 0 \\ 1 \\ 0 \end{bmatrix}, \begin{bmatrix} 0 \\ 0 \\ i \end{bmatrix} \right\}$ is a basis. Find the coordinate vector for $\boldsymbol{v}=$ $\begin{bmatrix} 1-i \\ 1 \\ 1 \end{bmatrix}$

$[\boldsymbol{v}]_S =$ ____________

6. In the vector space of 2×2 matrices, $\mathbf{S} = \left\{ \begin{bmatrix} 1 & 2 \\ 1 & -2 \end{bmatrix}, \begin{bmatrix} 0 & 1 \\ 1 & 0 \end{bmatrix}, \begin{bmatrix} 0 & 2 \\ 3 & 1 \end{bmatrix}, \begin{bmatrix} 1 & 0 \\ -1 & 2 \end{bmatrix} \right\}$ is a basis. For $\boldsymbol{v}= \begin{bmatrix} 8 & -11 \\ -26 & 13 \end{bmatrix}$, find $[\boldsymbol{v}]_S$.

$[\boldsymbol{v}]_S =$ ____________

Section 7.3

Changing Coordinates

The problem that we want to address next is the relationship between $[\boldsymbol{v}]_S$ and $[\boldsymbol{v}]_T$ where **S** and **T** are two bases for the same vector space V. If we have coordinates relative to the **T**-basis we want to be able to easily convert them to coordinates relative to the **S**-basis. This can be accomplished by multiplying by **the transition matrix from the T-basis to the S-basis**.

The transition matrix $\boldsymbol{P}$ from the **T**-basis to the **S**-basis has columns which are coordinates of the **T**-basis vectors relative to the **S**-basis.

Let **S** $= \{\boldsymbol{v_1},\boldsymbol{v_2}, \ldots\ \boldsymbol{v_n}\}$ and **T** $= \{\boldsymbol{w_1},\boldsymbol{w_2}, \ldots\ \boldsymbol{w_n}\}$ be bases for the same vector space V. The transition matrix $\boldsymbol{P}$ from the **T**-basis to the **S**-basis is given by

$$\boldsymbol{P} = [[\boldsymbol{w_1}]_S\ [\boldsymbol{w_2}]_S\ \ldots [\boldsymbol{w_n}]_S]$$

The following example illustrates how to compute the transition matrix (sometimes called the *change of basis matrix*) using MATLAB .

Example 1. In R^3 let $\mathbf{S} = \{\boldsymbol{v_1},\boldsymbol{v_2},\boldsymbol{v_3}\} = \left\{ \begin{bmatrix} 1 \\ 1 \\ 0 \end{bmatrix}, \begin{bmatrix} 1 \\ 0 \\ 1 \end{bmatrix}, \begin{bmatrix} 1 \\ 1 \\ 1 \end{bmatrix} \right\}$ and $\mathbf{T} = \{\boldsymbol{w_1},\boldsymbol{w_2},\boldsymbol{w_3}\} = \left\{ \begin{bmatrix} 1 \\ 2 \\ 1 \end{bmatrix}, \begin{bmatrix} 1 \\ 2 \\ 0 \end{bmatrix}, \begin{bmatrix} 1 \\ 0 \\ 2 \end{bmatrix} \right\}$ be bases. Find the transition matrix from the **T**-basis to the **S**-basis.

We proceed by finding the coordinates of the $\boldsymbol{w_i}$ relative to the **S**-basis. This computation requires that we solve a linear system whose coefficient matrix is composed of the vectors from the **S**-basis. Using the ideas for solving linear systems in MATLAB which we developed previously we have for $A = \begin{bmatrix} 1 & 1 & 1 \\ 1 & 0 & 1 \\ 0 & 1 & 1 \end{bmatrix}$

$$[\boldsymbol{w_1}]_S = A\backslash \begin{bmatrix} 1 & 2 & 1 \end{bmatrix}' \text{ which gives } \begin{bmatrix} 0 & -1 & 2 \end{bmatrix}'$$

$$[\boldsymbol{w_2}]_S = A\backslash \begin{bmatrix} 1 & 2 & 0 \end{bmatrix}' \text{ which gives } \begin{bmatrix} 1 & -1 & 1 \end{bmatrix}'$$

$$[\boldsymbol{w_3}]_S = A\backslash \begin{bmatrix} 1 & 0 & 2 \end{bmatrix}' \text{ which gives } \begin{bmatrix} -1 & 1 & 1 \end{bmatrix}'$$

Thus the transition matrix is $\boldsymbol{P} = \begin{bmatrix} 0 & 1 & -1 \\ -1 & -1 & 1 \\ 2 & 1 & 1 \end{bmatrix}$. For the vector $\boldsymbol{w} = \begin{bmatrix} 6 \\ 10 \\ 4 \end{bmatrix}$ it can be shown that $[\boldsymbol{w}]_T = \begin{bmatrix} 2 \\ 3 \\ 1 \end{bmatrix}$. Then the coordinates of $\boldsymbol{w}$ with respect to the **S**-basis are

$$[\boldsymbol{w}]_S = \boldsymbol{P} * [\boldsymbol{w}]_T = \begin{bmatrix} 2 \\ -4 \\ 8 \end{bmatrix}$$

This can be checked by direct calculation using the command **A\w**.

When working with the transition matrix $\boldsymbol{P}$ it is useful to have a technique for obtaining $\boldsymbol{P}$ directly. The method shown in Example 1 suggests one such approach. Instead of finding the coordinates of the individual vectors $\boldsymbol{w_1}$, $\boldsymbol{w_2}$, and $\boldsymbol{w_3}$ with respect to basis **S**, form the matrix $\boldsymbol{B}$ corresponding to the vectors in basis **T**, $\boldsymbol{B} = [\boldsymbol{w_1 w_2 w_3}]$, and then compute the transition matrix $\boldsymbol{P}$ from the MATLAB command **P** = **A\B**. A third method is to use MATLAB commands **M** = **[A B]**, **R** = **rref(M)**, **P** = **R(:,4:6)**. This method also produces the transition matrix $\boldsymbol{P}$ in Example 1.

Exercises 7.3

1. In $V = R^3$ let $\mathbf{S} = \{\boldsymbol{v_1}, \boldsymbol{v_2}, \boldsymbol{v_3}\} = \left\{ \begin{bmatrix} 1 \\ 1 \\ -1 \end{bmatrix}, \begin{bmatrix} 1 \\ 2 \\ 1 \end{bmatrix}, \begin{bmatrix} -1 \\ 1 \\ 0 \end{bmatrix} \right\}$ and $\mathbf{T} = \{\boldsymbol{w_1}, \boldsymbol{w_2}, \boldsymbol{w_3}\} = \left\{ \begin{bmatrix} 1 \\ 2 \\ 3 \end{bmatrix}, \begin{bmatrix} 3 \\ 1 \\ 2 \end{bmatrix}, \begin{bmatrix} 2 \\ 1 \\ 3 \end{bmatrix} \right\}$ be bases. Find the transition matrix from the **T**-basis to the **S**-basis. Name the transition matrix $\boldsymbol{P1}$. Record $\boldsymbol{P1}$ here.

2. Using the bases in Exercise 1, find the transition matrix from the **S**-basis to the **T**-basis. Call it ***P2***. Record ***P2*** here. What is the relationship between ***P1*** and ***P2***? (Hint: compute ***P1********P2***.)

3. Let V be the vector space of row vectors with three real entries and let

$$\mathbf{S} = \{\boldsymbol{v}_1,\boldsymbol{v}_2,\boldsymbol{v}_3\} = \left\{\begin{bmatrix} 1 & 0 & 1 \end{bmatrix}, \begin{bmatrix} 1 & 1 & 2 \end{bmatrix}, \begin{bmatrix} -1 & 1 & 2 \end{bmatrix}\right\}$$

and

$$\mathbf{T} = \{\boldsymbol{w}_1,\boldsymbol{w}_2,\boldsymbol{w}_3\} = \left\{\begin{bmatrix} 2 & 1 & 1 \end{bmatrix}, \begin{bmatrix} 2 & 2 & 1 \end{bmatrix}, \begin{bmatrix} 2 & 2 & 2 \end{bmatrix}\right\}$$

be bases. Find the transition matrix from the **T**-basis to the **S**-basis. Name the transition matrix ***P***. Record ***P*** here.

4. In the vector space of 2×2 matrices,

$$\mathbf{T} = \left\{\begin{bmatrix} 1 & 0 \\ -1 & 2 \end{bmatrix}, \begin{bmatrix} 0 & -8 \\ -12 & -4 \end{bmatrix}, \begin{bmatrix} 0 & -1 \\ -1 & 0 \end{bmatrix}, \begin{bmatrix} 0 & 4 \\ 1 & -4 \end{bmatrix}\right\} \text{ is a basis.}$$

Find the transition matrix from the **T**-basis to the **S**-basis given in Exercise 6 in Section 7.2.

<< NOTES; COMMENTS; IDEAS >>

LAB 8

The Determinant Function

Topics: the determinant function; its properties with respect to row operations; its behavior on nonsingular and singular matrices; a method to compute the value of the determinant function using row operations.

Introduction

We assume that the hand computation of determinants of 2×2 and 3×3 matrices has been discussed using the standard scheme involving products of diagonal arrangements of matrix entries. Figure 1 illustrates the so called 2×2 and 3×3 trick for determinants.

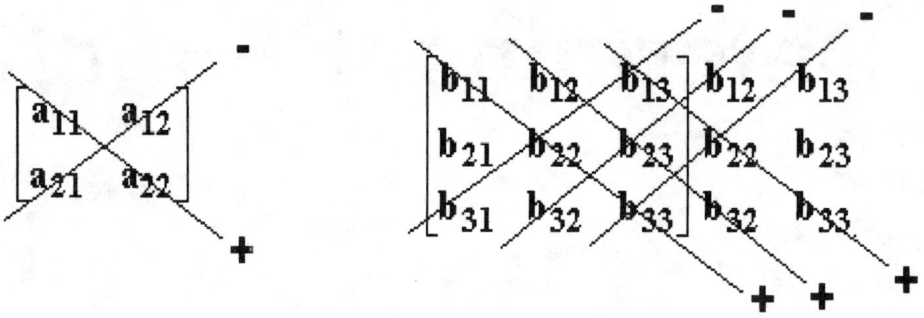

Figure 1.

The products of the entries are computed along the lines on the figure. The products are then added with the sign at the end of the line attached.

Section 8.1

Experiments to Investigate the Determinant Function

The *determinant* is a function from the square matrices of any size to the complex numbers. The function is denoted by **det**. Its value when acting on a square matrix $\boldsymbol{A}$ is denoted by **det** $\boldsymbol{A}$ when we write it and by the command **det(*A*)** in MATLAB . Some books also use the notation $| \boldsymbol{A} |$ for the value of the determinant. We will use MATLAB to explore the properties of the determinant function by a series of experiments.

Exercises 8.1

1. Construct a 2×2 matrix with a row of zeros and compute its determinant. Repeat for a 3×3 and a 4×4 matrix with a row of zeros. Construct 2×2, 3×3, and 4×4 matrices each with a column of zeros and compute their determinants.

> Conjecture:
>
> The determinant of a matrix with a row or column of zeros is ____________ .

2. Construct a 2×2 matrix with two equal rows and compute its determinant. Repeat the computation for a 3×3 and a 4×4 matrix. Construct 2×2, 3×3, and a 4×4 matrices with two equal columns and compute their determinants.

> Conjecture:
>
> The determinant of a matrix with two equal rows or columns is ____________ .

3. The MATLAB command **A = fix(10*rand(3))** generates a 3×3 real matrix A. Compute **det(A)** and **det(A′)**. Change the 3 in the MATLAB command to several other integer values like 2, 4, 5, 6 and repeat the computations.

> Conjecture:
>
> The determinant of a real matrix and the determinant of its transpose are ____________ .

4. Construct a 2×2 diagonal matrix with diagonal entries 5 and 3 and record the value of its determinant. ____________

Construct a 2×2 diagonal matrix with diagonal entries -2 and 9 and record the value of its determinant. ____________

Construct a 3×3 diagonal matrix with diagonal entries 2, 7, and -1 and record the value of its determinant. ____________

Construct a 3×3 diagonal matrix with diagonal entries 4, 0, and 3 and record the value of its determinant. ____________

> Conjecture:
>
> The determinant of a diagonal matrix is ______________________.

In MATLAB we can construct a 3×3 upper triangular matrix using the command **A = triu(fix(10*rand(3)))**. Generate several upper triangular matrices of size 3 and compute their determinants. Repeat the procedure for several other sizes.

> Conjecture:
>
> The determinant of an upper triangular matrix is ______________________.

5. Let $\boldsymbol{A} = \begin{bmatrix} 1 & 2 & 3 \\ 4 & 5 & 6 \\ 7 & 8 & 0 \end{bmatrix}$. Compute and record **det(A)** = ____________.

We will perform a series of row operations on $\boldsymbol{A}$ and compute the determinant of each new matrix. Always perform the row operation on the original matrix $\boldsymbol{A}$.

> Notation:
>
> - $\boldsymbol{A}_{R_i \leftrightarrow R_j}$ means interchange row i with row j in matrix $\boldsymbol{A}$.
> - $\boldsymbol{A}_{kR_i + R_j}$ means replace row j of $\boldsymbol{A}$ by k times row i plus row j.
> - $\boldsymbol{A}_{kR_i}$ means multiply row i of matrix $\boldsymbol{A}$ by scalar k.

Let $\boldsymbol{B} = \boldsymbol{A}_{R_1 \leftrightarrow R_2}$; **det(B)** = ____________
How is **det(B)** related to **det(A)**? ______________________

Let $\boldsymbol{C} = \boldsymbol{A}_{R_2 \leftrightarrow R_3}$; **det(C)** = ____________
How is **det(C)** related to **det(A)**? ______________________

Let $\boldsymbol{D} = \boldsymbol{A}_{2R_1 + R_2}$; **det(D)** = ____________
How is **det(D)** related to **det(A)**? ______________________

Let $\boldsymbol{E} = \boldsymbol{A}_{-4R_2 + R_3}$; **det(E)** = ____________
How is **det(E)** related to **det(A)**? ______________________

Let $F = A_{3R_1}$; $\mathbf{det(F)}$ = ____________
How is $\mathbf{det(F)}$ related to $\mathbf{det(A)}$? ____________________

Let $G = A_{-2R_2}$; $\mathbf{det(G)}$ = ____________
How is $\mathbf{det(G)}$ related to $\mathbf{det(A)}$? ____________________

Let $H = A_{1/2R_3}$; $\mathbf{det(H)}$ = ____________
How is $\mathbf{det(H)}$ related to $\mathbf{det(A)}$? ____________________

If you have difficulty filling in the following responses repeat the previous experiment with

$$A = \begin{bmatrix} 2 & -5 & 3 \\ 0 & 2 & -1 \\ 3 & 2 & 1 \end{bmatrix}.$$

Conjectures:

If we interchange rows the determinant ____________________.
If we replace one row by a linear combination of itself with another row the determinant ____________________.
If we multiply a row by scalar k the determinant ____________________.

6. Fill in the blanks.

a) Let $A = \begin{bmatrix} 1 & 2 & 3 \\ 4 & 5 & 6 \\ 7 & 8 & 9 \end{bmatrix}$; $\mathbf{rref(A)}$ = ____________ $\mathbf{det(A)}$ = ______

$\mathbf{det(rref(A))}$ = ______

b) Let $B = \begin{bmatrix} 1 & 2 \\ 2 & 4 \end{bmatrix}$; $\mathbf{rref(B)}$ = ____________ $\mathbf{det(B)}$ = ______

$\mathbf{det(rref(B))}$ = ______

c) Let $C = \begin{bmatrix} 1 & 1 & 1 \\ 2 & 1 & -1 \\ 3 & 2 & 0 \end{bmatrix}$; $\mathbf{rref(C)}$ = ____________ $\mathbf{det(C)}$ = ______

$\mathbf{det(rref(C))}$ = ______

d) Let $D = \begin{bmatrix} 2 & 1 & 0 \\ 1 & 2 & 1 \\ 0 & 1 & 2 \end{bmatrix}$; $\mathbf{rref(D)}$ = ____________ $\mathbf{det(D)}$ = ______

$\mathbf{det(rref(D))}$ =______

e) <u>True or False:</u> For any square matrix $\mathbf{Q}$, $\mathbf{det(Q)} = \mathbf{det(rref(Q))}$. ____________

f) Based upon the few experiments in parts a) – d), does there seem to be a connection between the following:

rref is I	det is zero
rref is not I	det is not zero

Draw an arrow between those that appear to be related.

<u>Conjectures: Let **Q** be a square matrix.</u>

If $\mathbf{rref(Q)} = \mathbf{I}$, then $\mathbf{det(Q)}$ is ____________________ .
If $\mathbf{rref(Q)} \neq \mathbf{I}$, then $\mathbf{det(Q)}$ is ____________________ .
The determinant of a nonsingular matrix is ____________________ .
The determinant of a singular matrix is ____________________ .

7. A general way to compute a determinant is to use row operations to reduce it to upper triangular form, keeping track of how the row operations change its value, and then use the fact that the determinant of an upper triangular matrix is the product of the diagonal entries. (See Exercise 4.) To illustrate, let $A = \begin{bmatrix} 2 & 3 & 1 \\ 1 & -2 & 2 \\ 3 & 0 & 4 \end{bmatrix}$. The objective is to compute $\mathbf{det(A)}$ using properties of the determinant.

Let $B = A_{R_1 \leftrightarrow R_2} = \begin{bmatrix} 1 & -2 & 2 \\ 2 & 3 & 1 \\ 3 & 0 & 4 \end{bmatrix}$;

$\mathbf{det(B)} = \boxed{}\ \mathbf{det(A)} \Longrightarrow \mathbf{det(A)} = \boxed{}\ \mathbf{det(B)}$

Let $C = B_{-2R_1+R_2} = \begin{bmatrix} 1 & -2 & 2 \\ 0 & 7 & -3 \\ 3 & 0 & 4 \end{bmatrix}$;

$\mathbf{det(C)} = \boxed{}\ \mathbf{det(B)} \Longrightarrow \mathbf{det(A)} = \boxed{}\ \mathbf{det(C)}$

Let $\boldsymbol{D} = \boldsymbol{C}_{-3R_1+R_3} = \begin{bmatrix} 1 & -2 & 2 \\ 0 & 7 & -3 \\ 0 & 6 & -2 \end{bmatrix}$;

$\mathbf{det(D)} = \boxed{}\ \mathbf{det(C)} \Longrightarrow \mathbf{det(A)} = \boxed{}\ \mathbf{det(D)}$

Let $\boldsymbol{E} = \boldsymbol{D}_{1/7R_2} = \begin{bmatrix} 1 & -2 & 2 \\ 0 & 1 & -3/7 \\ 0 & 6 & -2 \end{bmatrix}$;

$\mathbf{det(E)} = \boxed{}\ \mathbf{det(D)} \Longrightarrow \mathbf{det(A)} = \boxed{}\ \mathbf{det(E)}$

Let $\boldsymbol{F} = \boldsymbol{E}_{-6R_2+R_3} = \begin{bmatrix} 1 & -2 & 2 \\ 0 & 1 & -3/7 \\ 0 & 0 & 4/7 \end{bmatrix}$;

$\mathbf{det(F)} = \boxed{}\ \mathbf{det(E)} \Longrightarrow \mathbf{det(A)} = \boxed{}\ \mathbf{det(F)}$

Now compute $\mathbf{det(A)}$ from $\mathbf{det(F)}$. ____________________
Check your work by computing $\mathbf{det(A)}$ directly.

8. Follow the procedure in Exercise 7 to compute $\mathbf{det(A)}$ where $\boldsymbol{A} = \begin{bmatrix} 5 & 1 & 0 \\ 0 & 2 & 1 \\ -1 & 3 & 1 \end{bmatrix}$. Show your work below.

<< NOTES; COMMENTS; IDEAS >>

LAB 9

Inner Product Spaces

Topics: vectors in MATLAB ; standard inner product or dot product in R^n; command **dot**; norm or length of a vector; command **norm**; angle between vectors; orthogonal vectors; unit vectors; routine **uball**.

Introduction

This lab introduces measurements: length, distance, and angle. Section 9.1 introduces the dot product (inner product) of vectors. The definition of an inner product space is examined in the exercises. Section 9.2 uses the dot product to determine the length (or norm) of a vector and to define the distance between a pair of vectors. Section 9.3 uses dot products and the norm to determine the angle between vectors. The exercises consider the Cauchy-Schwarz inequality in an experimental fashion and the norm on an abstract vector space. Other exercises provide a geometric visualization of a family of norms on R^2 and R^3 using routine **uball**.

Section 9.1

The Standard Inner Product

In MATLAB an n-vector can be considered either a column or a row with n entries. We denote the vector space of n-vectors by R^n. We sometimes use the term vector for an element of R^n. The context will tell us whether we are to consider a vector as a row or column. R^n is an n-dimensional vector space and various inner products can be defined on it. Here our primary concern is with a particular inner product, called the *standard inner product* or *dot product*, on R^n. For a pair of vectors $\boldsymbol{u}$ and $\boldsymbol{v}$ in R^n the standard inner product of $\boldsymbol{u}$ and $\boldsymbol{v}$ is computed as

$$\sum_{j=1}^{n} u_j v_j = u_1 v_1 + u_2 v_2 + \cdots + u_n v_n$$

To denote the inner product of $\boldsymbol{u}$ and $\boldsymbol{v}$ we use

$$(\boldsymbol{u},\boldsymbol{v}) \qquad \text{or} \qquad \boldsymbol{u} \cdot \boldsymbol{v}$$

In MATLAB , once vectors $\boldsymbol{u}$ and $\boldsymbol{v}$ have been entered we use the command

dot(u,v)

to compute their dot product. If $\boldsymbol{u}$ and $\boldsymbol{v}$ are not the same size an error message is displayed. If $\boldsymbol{u}$ and $\boldsymbol{v}$ are both real column vectors (of the same size) in MATLAB then we can compute the standard inner product directly as the matrix product **u′∗v** or equivalently as **v′∗u**. The inner product of complex vectors is treated in Exercise 3.

Example 1. Let $\boldsymbol{u} = \begin{bmatrix} 3 \\ -1 \\ 0 \\ 2 \end{bmatrix}$ and $\boldsymbol{v} = \begin{bmatrix} 0 \\ 2 \\ 1 \\ -4 \end{bmatrix}$ be vectors in R^4. Enter these vectors as columns into MATLAB . Then we have that

dot(u,v)

displays

```
ans =

   -10
```

Verify in MATLAB that the matrix products **u′∗v** and **v′∗u** give the same result.

A vector space with an inner product defined on it is called an **inner product space**.

There is a wide variety of inner product spaces that play an important role in applications. Inner products provide a convenient tool used to define other types of measures on elements of vector spaces. See Sections 9.2 and 9.3.

Exercises 9.1

1. Compute the dot product of each of the following pairs of vectors. Record the result in the space provided.

a) $\boldsymbol{v} = \begin{bmatrix} -1 \\ 3 \\ 2 \end{bmatrix}, \boldsymbol{w} = \begin{bmatrix} 2 \\ -1 \\ 0 \end{bmatrix}$. ______ **b)** $\boldsymbol{v} = \begin{bmatrix} 2 \\ -3 \\ 0 \\ -2 \end{bmatrix}, \boldsymbol{w} = \begin{bmatrix} -1 \\ 2 \\ 4 \\ -4 \end{bmatrix}$. ______

c) $\boldsymbol{v} = \begin{bmatrix} 0 \\ -1 \\ 3 \\ 4 \end{bmatrix}, \boldsymbol{w} = \begin{bmatrix} -1 \\ 5 \\ 6 \\ -3 \end{bmatrix}$. ______ **d)** $\boldsymbol{v} = \begin{bmatrix} 3 \\ 1 \\ 3 \\ -1 \end{bmatrix}, \boldsymbol{w} = \begin{bmatrix} 4 \\ -3 \\ 2 \\ 2 \end{bmatrix}$. ______

2. Let **V** be the vector space of 2×3 matrices. In Examples 3 and 4 of Section 6.1 we showed how to use the MATLAB command **reshape** to associate a column (in this case with six entries) with an element of **V**. When MATLAB command **dot** detects input which is not rows or columns of the same size it uses the **reshape** command to automatically make the column association. The command for the dot product of two matrices $\boldsymbol{A}$ and $\boldsymbol{B}$ in **V** is:

dot(A,B)

The description above applies to any size matrix. The only restriction is that **dot** requires the matrices to be the same size. Compute the dot product of the following pairs of matrices. Record the result in the space provided. (Note: The association of a column with an $m \times n$ matrix is an example of a function between vector spaces known as an *isomorphism.*)

a) $\boldsymbol{A} = \begin{bmatrix} 0 & 3 \\ -1 & 4 \end{bmatrix}, \boldsymbol{B} = \begin{bmatrix} -1 & 6 \\ 5 & -3 \end{bmatrix}$. ____________

(Use the **reshape** command on matrices $\boldsymbol{A}$ and $\boldsymbol{B}$ to determine the associated columns and compute the dot product of the columns directly. Compare your results with Exercise 1c).

b) $\boldsymbol{A} = \begin{bmatrix} 3 & 3 \\ 1 & -1 \end{bmatrix}, \boldsymbol{B} = \begin{bmatrix} 4 & 2 \\ -3 & 2 \end{bmatrix}$. ____________

(Use the **reshape** command on matrices $\boldsymbol{A}$ and $\boldsymbol{B}$ to determine the associated columns and compute the dot product of the columns directly. Compare your results with Exercise 1d).

c) $\boldsymbol{A} = \begin{bmatrix} 2 & -1 & 4 \\ 0 & 2 & 7 \\ 4 & 5 & 1 \end{bmatrix}, \boldsymbol{B} = \begin{bmatrix} 0 & 2 & -4 \\ 6 & 5 & -4 \\ 1 & -1 & -3 \end{bmatrix}$. ____________

d) $\boldsymbol{A} = \begin{bmatrix} 1 & 2 & 3 & 4 \\ -3 & 4 & 0 & 1 \end{bmatrix}, \boldsymbol{B} = \begin{bmatrix} 1 & 2 & 2 & 1 \\ -1 & 3 & 3 & -1 \end{bmatrix}$. ____________

3. In Exercises 2.1 we discussed that the prime (′) operator in MATLAB returns the *conjugate transpose* of a complex matrix. Let $\mathbf{C}^n$ be the vector space of columns with n complex entries. Then the standard inner product of vectors $\boldsymbol{x}$ and $\boldsymbol{y}$ in $\mathbf{C}^n$ is defined as

$$(\boldsymbol{x}, \boldsymbol{y}) = (\text{conjugate transpose of } \boldsymbol{x}) * \boldsymbol{y}$$

In MATLAB we compute this directly as $\mathbf{x}' * \mathbf{y}$ or using command **dot(x,y)**. Compute the inner product of each of the following pairs of complex vectors. Record your result in the space provided. (Note: sometimes the complex inner product is defined as $(\boldsymbol{x}, \boldsymbol{y}) =$ (conjugate transpose of $\boldsymbol{y}$) $*\boldsymbol{x}$, or $\boldsymbol{y}' * \boldsymbol{x}$.)

a) $\boldsymbol{x} = \begin{bmatrix} 1 \\ i \end{bmatrix}, \boldsymbol{y} = \begin{bmatrix} 2+i \\ 3-4i \end{bmatrix}$. ____________

b) $\boldsymbol{x} = \begin{bmatrix} -1+i \\ 2i \\ 3-4i \end{bmatrix}, \boldsymbol{y} = \begin{bmatrix} 2+3i \\ 4 \\ 1-2i \end{bmatrix}$. ____________

c) $x = \begin{bmatrix} 1+i \\ 3i \\ 0 \end{bmatrix}$, $y = \begin{bmatrix} -2+3i \\ 2-2i \\ 1-2i \end{bmatrix}$. ____________

4. An inner product is a function F from a vector space $\mathbf{V}$ to the real numbers or the complex numbers, depending upon the scalars permitted in the vector space, that satisfies the following properties:

(i) For any vector $\boldsymbol{v}$ in $\mathbf{V}$, $F(\boldsymbol{v}, \boldsymbol{v}) \geq 0$ and equality holds if and only if $\boldsymbol{v}$ is the zero vector in $\mathbf{V}$.

(ii) For any $\boldsymbol{v}, \boldsymbol{w}$ in $\mathbf{V}$, $F(\boldsymbol{v}, \boldsymbol{w}) = \overline{F(\boldsymbol{w}, \boldsymbol{v})}$. (Here the bar means take the conjugate.)

(iii) For any $\boldsymbol{v}, \boldsymbol{w}$, and $\boldsymbol{u}$ in $\mathbf{V}$, $F(\boldsymbol{v} + \boldsymbol{w}, \boldsymbol{u}) = F(\boldsymbol{v}, \boldsymbol{u}) + F(\boldsymbol{w}, \boldsymbol{u})$.

(iv) For any $\boldsymbol{v}, \boldsymbol{w}$ in $\mathbf{V}$ and any scalar k, $F(k\boldsymbol{v}, \boldsymbol{w}) = \bar{k}F(\boldsymbol{v}, \boldsymbol{w})$.

Use MATLAB to show that each of the following functions is **not** an inner product. State a property listed above that is violated.

a) Let $\mathbf{V}$ be the vector space R^3. For $\boldsymbol{v} = [\boldsymbol{v_1 v_2 v_3}]^T$ and $\boldsymbol{w} = [\boldsymbol{w_1 w_2 w_3}]^T$, define

$$F(\boldsymbol{v}, \boldsymbol{w}) = |v_1| \cdot |w_1| + |v_2| \cdot |w_2| + |v_3| \cdot |w_3|$$

In MATLAB function F can be computed by the command **abs(v)'*abs(w)**.

b) Let $\mathbf{V}$ be the vector space C^2. For $\boldsymbol{v} = \begin{bmatrix} v_1 \\ v_2 \end{bmatrix}$ and $\boldsymbol{w} = \begin{bmatrix} w_1 \\ w_2 \end{bmatrix}$, define

$$F(\boldsymbol{v}, \boldsymbol{w}) = v_1 w_1 + v_2 w_2$$

In MATLAB function F can be computed by the command **v.'*w**. (Note that operation .' is not the same as ', the conjugate transpose operator.)

c) Let $\mathbf{V}$ be the vector space R^2. Define

$$F(\boldsymbol{v}, \boldsymbol{w}) = \boldsymbol{v}^T C \boldsymbol{w}$$

where $C = \begin{bmatrix} 1 & 2 \\ 2 & 1 \end{bmatrix}$. (Hint: Find a nonzero vector $\boldsymbol{v}$ so that $\boldsymbol{Cv} = (-1)\boldsymbol{v}$. To do this solve the homogeneous system $(\boldsymbol{C} + \boldsymbol{I_2})\boldsymbol{v} = \boldsymbol{0}$. Then compute $F(\boldsymbol{v}, \boldsymbol{v})$.)

d) Let **V** be the vector space R^2. Define

$$F(\boldsymbol{v}, \boldsymbol{w}) = \boldsymbol{v^T C w}$$

where $C = \begin{bmatrix} 1 & 1 \\ 2 & 2 \end{bmatrix}$. (Hint: Find a nonzero vector $\boldsymbol{v}$ in the null space of $\boldsymbol{C}$ and compute $F(\boldsymbol{v}, \boldsymbol{v})$.)

Section 9.2

Length and Distance

Once an inner product $(\boldsymbol{u}, \boldsymbol{v})$ or $\boldsymbol{u} \cdot \boldsymbol{v}$ has been defined on a vector space **V** we call **V** an inner product space. In an inner product space we use the inner product function to define other concepts. The *length* or *norm* of a vector $\boldsymbol{v}$ is denoted $\|\boldsymbol{v}\|$ and is defined by

$$\|\boldsymbol{v}\| = \sqrt{(\boldsymbol{v}, \boldsymbol{v})} = \sqrt{\boldsymbol{v} \cdot \boldsymbol{v}}$$

The notion of a norm is well defined since $(\boldsymbol{v}, \boldsymbol{v}) \geq 0$ for any inner product, with equality holding if and only if $\boldsymbol{v}$ is the zero vector.

For the standard inner product of R^n with $\boldsymbol{v} = \begin{bmatrix} v_1 \\ v_2 \\ \vdots \\ v_n \end{bmatrix}$ we have

$$\|\boldsymbol{v}\| = \sqrt{\sum_{j=1}^{n} v_j^2} = \sqrt{v_1^2 + v_2^2 + \cdots + v_n^2}$$

In MATLAB we can compute the norm based on the standard inner product on R^n (or C^n; see Exercise 3 in Section 9.1) using the **norm** command as in

norm(v)

or directly as

sqrt(v′∗v)

Example 1. Let $\boldsymbol{u} = \begin{bmatrix} 3 \\ -1 \\ 0 \\ 2 \end{bmatrix}$ and $\boldsymbol{v} = \begin{bmatrix} 0 \\ 2 \\ 1 \\ -4 \end{bmatrix}$ be vectors in R^4. Enter these vectors as columns into MATLAB . Consider R^4 with the standard inner product and compute the length of each vector. Commands

ul = norm(u), vl = norm(v)

display

```
ul =

    3.7417

vl =

    4.5826
```

The *distance* between two vectors $\boldsymbol{u}$ and $\boldsymbol{v}$ in an inner product space $\mathbf{V}$ is defined as the norm of their difference, $\|\boldsymbol{u}-\boldsymbol{v}\|$. In terms of the inner product, $\|\boldsymbol{u}-\boldsymbol{v}\| = \sqrt{(\boldsymbol{u}-\boldsymbol{v}, \boldsymbol{u}-\boldsymbol{v})}$.

For the standard inner product on R^n, with $\boldsymbol{u} = \begin{bmatrix} u_1 \\ u_2 \\ \vdots \\ u_n \end{bmatrix}$ and $\boldsymbol{v} = \begin{bmatrix} v_1 \\ v_2 \\ \vdots \\ v_n \end{bmatrix}$,

$$\|\boldsymbol{u}-\boldsymbol{v}\| = \sqrt{\textstyle\sum_{j=1}^{n}(u_j - v_j)^2} = \sqrt{(u_1-v_1)^2 + (u_2-v_2)^2 + \cdots + (u_n-v_n)^2}$$

In MATLAB we compute $\|\boldsymbol{u}-\boldsymbol{v}\|$ using command **norm(u − v)**.

Example 2. Let $\boldsymbol{u} = \begin{bmatrix} 3 \\ -1 \\ 0 \\ 2 \end{bmatrix}$ and $\boldsymbol{v} = \begin{bmatrix} 0 \\ 2 \\ 1 \\ -4 \end{bmatrix}$ be vectors in R^4. Enter these vectors as columns into MATLAB . Consider R^4 with the standard inner product and compute the distance between $\boldsymbol{u}$ and $\boldsymbol{v}$. The command

d = norm(u − v)

displays

```
d =

    7.4162
```

Example 3. Let **V** be the vector space of 2×2 matrices. For $\boldsymbol{A}$ and $\boldsymbol{B}$ in **V**, define their inner product as follows in MATLAB

$(\boldsymbol{A}, \boldsymbol{B})$ = **dot(A,B)**

as in Exercise 2 in Section 9.1. With this function **V** is an inner product space, hence we can compute the length of a 2×2 matrix and the distance between 2×2 matrices. Here the length of matrix $\boldsymbol{A}$, denoted by $\|\boldsymbol{A}\|$, is computed in MATLAB as

norm(reshape(A,4,1)) or **sqrt(dot(A,A))**

and the distance between matrices $\boldsymbol{A}$ and $\boldsymbol{B}$, $\|\boldsymbol{A} - \boldsymbol{B}\|$, is computed in MATLAB as

norm(reshape(A,4,1) - reshape(B,4,1)) or **sqrt (dot(A-B,A-B))**

For $\boldsymbol{A} = \begin{bmatrix} 1 & 2 \\ 2 & 1 \end{bmatrix}$ and $\boldsymbol{B} = \begin{bmatrix} 3 & -2 \\ 0 & 4 \end{bmatrix}$ verify in MATLAB that $\|\boldsymbol{A}\| = \sqrt{10}$, $\|\boldsymbol{B}\| = \sqrt{29}$, and $\|\boldsymbol{A} - \boldsymbol{B}\| = \sqrt{33}$. (MATLAB displays a decimal expression in place of the square roots above. Use command **ans**^2 to verify each result.)

WARNING: The command **norm(A)** has a different meaning than $\|\boldsymbol{A}\|$ used in this example. Execute command **norm(A)** and you will see that it produces a value of 3 and not $\sqrt{3}$. Type **help norm**. You will see that the command refers to the largest singular value of matrix $\boldsymbol{A}$, a concept beyond the scope of this book. The behavior of the **norm** command on vectors is discussed in Example 2.

Exercises 9.2

1. Let $\mathbf{V} = R^5$ with the standard inner product. For vectors

$$\boldsymbol{v} = \begin{bmatrix} 2 \\ 1 \\ 2 \\ 1 \\ 2 \end{bmatrix}, \quad \boldsymbol{u} = \begin{bmatrix} 1 \\ 2 \\ 3 \\ 4 \\ 5 \end{bmatrix}, \text{ and } \boldsymbol{w} = \begin{bmatrix} 0 \\ 2 \\ 0 \\ 1 \\ 3 \end{bmatrix}$$

compute the following in MATLAB and display the results in the space provided.

a) length of $\boldsymbol{v}$ ____________ **b)** length of $\boldsymbol{u}$ ____________

c) length of $\boldsymbol{w}$ ____________ **d)** distance from $\boldsymbol{v}$ to $\boldsymbol{u}$ ____________

e) distance from $\boldsymbol{v}$ to $\boldsymbol{w}$ ____________ **f)** distance from $\boldsymbol{u}$ to $\boldsymbol{w}$ ____________

2. a) Compute **dot(v,v)** and **norm(v)** for the vector $\boldsymbol{v}$ in Exercise 1. How are they related?

____

b) Repeat part a) for the vectors **u** and **w** from Exercise 1. Based on this limited numeric evidence, form a conjecture relating **dot(x,x)** and **norm(x)** for any vector $\boldsymbol{x}$.

____

3. Let **V** be the vector space of 4×2 matrices with real entries with the inner product as given in Example 3. (Note: use **reshape(A,8,1)** here.) For matrices

$$\boldsymbol{A} = \begin{bmatrix} 1 & 2 \\ 1 & 2 \\ 2 & 1 \\ 2 & 1 \end{bmatrix}, \quad \boldsymbol{B} = \begin{bmatrix} 1 & 0 \\ 0 & 1 \\ 1 & 0 \\ 0 & 1 \end{bmatrix}, \text{ and } \boldsymbol{C} = \begin{bmatrix} 2 & 1 \\ 3 & 4 \\ 1 & 2 \\ 3 & 4 \end{bmatrix}$$

compute the following in MATLAB and display the results in the space provided.

a) $\|\boldsymbol{A}\|$, ____________ **b)** $\|\boldsymbol{B}\|$ ____________

c) $\|\boldsymbol{C}\|$ ____________ **d)** $\|\boldsymbol{A} - \boldsymbol{B}\|$ ____________

e) $\|\boldsymbol{A} - \boldsymbol{C}\|$ ____________ **f)** $\|\boldsymbol{B} - \boldsymbol{C}\|$ ____________

4. A *unit vector* is a vector of length one. For a nonzero vector $\boldsymbol{v}$ in an inner product space $\mathbf{V}$, a unit vector in the same direction (see Section 9.3) as $\boldsymbol{v}$ is given by

$$\boldsymbol{v}/\ \|\boldsymbol{v}\|$$

a) Find a unit vector corresponding to vector $\boldsymbol{v}$ in Exercise 1. Record your result below.

b) Find a unit vector corresponding to vector $\boldsymbol{w}$ in Exercise 1. Record your result below.

c) Find a unit vector corresponding to vector $\boldsymbol{A}$ in Exercise 3. Record your result below.

5. Let $\mathbf{V} = C^3$, the vector space of columns with 3 complex entries, with the standard inner product as defined in Exercise 3 in Section 9.1. For vectors

$$\boldsymbol{v} = \begin{bmatrix} 2+i \\ 3-4i \\ 2 \end{bmatrix}, \ \boldsymbol{u} = \begin{bmatrix} i \\ 4+2i \\ 1-i \end{bmatrix}, \text{ and } \boldsymbol{w} = \begin{bmatrix} 1+i \\ 0 \\ -2i \end{bmatrix}$$

compute the following in MATLAB .

a) length of $\boldsymbol{v}$ ____________ **b)** length of $\boldsymbol{u}$ ____________

c) length of $\boldsymbol{w}$ ____________ **d)** distance from $\boldsymbol{v}$ to $\boldsymbol{u}$ ____________

e) distance from $\boldsymbol{v}$ to $\boldsymbol{w}$ ____________ **f)** distance from $\boldsymbol{u}$ to $\boldsymbol{w}$ ____________

6. Let $\mathbf{V}$ be the inner product space R^n with the standard inner product. For pairs of vectors $\boldsymbol{v}$ and $\boldsymbol{w}$ in R^n compute the following quantities and record their values in the table below. Compute $\|\boldsymbol{v}\| * \|\boldsymbol{w}\|$ and $|(\boldsymbol{v}, \boldsymbol{w})|$. In MATLAB use commands **norm(v)*norm(w)**

and **abs(dot(v,w))**.

a) $v = \begin{bmatrix} 1 \\ 3 \\ 1 \end{bmatrix}$, $w = \begin{bmatrix} 2 \\ 1 \\ 4 \end{bmatrix}$ **b)** $v = \begin{bmatrix} 1 \\ 2 \\ -1 \\ -3 \end{bmatrix}$, $w = \begin{bmatrix} 4 \\ -1 \\ 0 \\ 2 \end{bmatrix}$ **c)** $v = \begin{bmatrix} 2 \\ 3 \end{bmatrix}$, $w = \begin{bmatrix} -6 \\ -9 \end{bmatrix}$

d) $v = \begin{bmatrix} 1 \\ 2 \end{bmatrix}$, $w = \begin{bmatrix} 3 \\ 4 \end{bmatrix}$ **e)** $v = \begin{bmatrix} 1 \\ 2 \\ -1 \\ 0 \\ 1 \end{bmatrix}$, $w = \begin{bmatrix} 5 \\ -1 \\ 2 \\ 1 \\ 0 \end{bmatrix}$ **f)** $v = \begin{bmatrix} 1 \\ 0 \\ 1 \end{bmatrix}$, $w = \begin{bmatrix} 0 \\ 3 \\ 0 \end{bmatrix}$

	norm(v) * norm(w) = $\| v \| * \| w \|$	**abs(dot(v,w))** = $\| (v, w) \|$
a)		
b)		
c)		
d)		
e)		
f)		

Based upon the experimental evidence in the preceding table state a conjecture about which value is greater, $\|v\| * \|w\|$ or $|(v, w)|$. Record your conjecture in the space below. Check your conjecture with other pairs of vectors in R^n.

Conjecture: ____________________

7. We have used an inner product to define the length or norm of vectors. However, a norm (function) can be defined directly on a vector space $\mathbf{V}$ as a function N from $\mathbf{V}$ to the real numbers that satisfies

(i) $N(\boldsymbol{v}) > 0$, for any nonzero vector in $\mathbf{V}$.

(ii) $N(\text{zero vector}) = 0$.

(iii) $N(k\boldsymbol{v}) = |k|N(\boldsymbol{v})$, for any scalar k and any vector $\boldsymbol{v}$.

(iv) $N(\boldsymbol{v} + \boldsymbol{u}) \leq N(\boldsymbol{v}) + N(\boldsymbol{u})$, for any vectors $\boldsymbol{v}$ and $\boldsymbol{u}$.

The norm defined using the standard inner product on R^n is called the 2-*norm* and for emphasis we write $\|\boldsymbol{v}\|_2$ for the 2−norm of vector $\boldsymbol{v}$. Two other norms on R^n that are simple and useful are the 1-*norm* and the *max norm*, sometimes called the ∞-*norm*. The 1-norm is denoted $\|\boldsymbol{v}\|_1$ and is defined as

$$\|\boldsymbol{v}\|_1 = \sum_{j=1}^{n} | v_j | = | v_1 | + | v_2 | + \cdots + | v_n | .$$

The ∞-norm is denoted $\|\boldsymbol{v}\|_\infty$ and is defined as

$$\|\boldsymbol{v}\|_\infty = \max\{| v_j |, j = 1, 2, \ldots, n\} = \max\{| v_1 |, | v_2 |, \ldots, | v_n |\}$$

In MATLAB the 2-norm is computed as before by command **norm(v)** or by **norm(v,2)**. The 1-norm is computed by command **norm(v,1)** and the ∞-norm by **norm(v,inf)**. Use MATLAB to compute the following and record the results in the space provided.

a) $\boldsymbol{v} = \begin{bmatrix} 1 \\ 1 \\ 1 \end{bmatrix}$ $\|\boldsymbol{v}\|_1$ ________ $\|\boldsymbol{v}\|_2$ ________ $\|\boldsymbol{v}\|_\infty$ ________

b) $\boldsymbol{v} = \begin{bmatrix} 2 \\ 3 \\ -1 \\ 4 \end{bmatrix}$ $\|\boldsymbol{v}\|_1$ ________ $\|\boldsymbol{v}\|_2$ ________ $\|\boldsymbol{v}\|_\infty$ ________

c) $\boldsymbol{v} = \begin{bmatrix} 1 \\ -1 \\ 2 \\ -2 \end{bmatrix}$ $\|\boldsymbol{v}\|_1$ ________ $\|\boldsymbol{v}\|_2$ ________ $\|\boldsymbol{v}\|_\infty$ ________

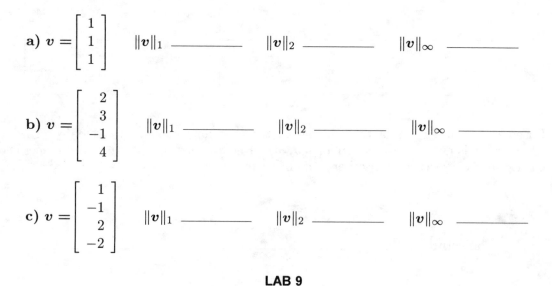

d) $v = \begin{bmatrix} 1 \\ 0 \\ 0 \end{bmatrix}$ $\|v\|_1$ ________ $\|v\|_2$ ________ $\|v\|_\infty$ ________

8. Since a norm defines how the length of a vector is measured, a norm determines a geometry for the vectors. One way to view the geometry imposed by different norms is to investigate the shape of the set of all vectors with norm equal to 1. Such sets are called **unit balls**. If we include the interior of these balls we say we are looking at the unit disk in a particular norm. In R^n the norm based on the standard inner product is computed as the square root of the sum of the squares of the components of the vector, hence we expect the **unit ball** or unit disk to be related to a circle in R^2 and a sphere in R^3.

The 2-norm is a particular case of a family of vector norms called *p-norms*. For a vector $\boldsymbol{x}$ in R^n the p-norm of $\mathbf{x}$ is denoted by $\| \boldsymbol{x} \|_p$ and computed by the expression

$$\| \boldsymbol{x} \|_p = \left(\sum_{i=1}^{n} |x_i|^p \right)^{1/p}$$

Each choice of $p \geq 1$ gives a different norm on R^n.

If we restrict our attention to R^2 and R^3 then we can have MATLAB display the shape of unit balls for various p-norms. The routine **uball** can be used to see the shape of unit balls. Type **uball** and then follow the screen directions for the following cases.

a) Choose the norm value as 2 and 500 trials. Describe the unit ball in the 2-norm in R^2.

____

In R^3.

____

b) Choose the norm value as 1 and 500 trials. Describe the unit ball in the 1-norm in R^2.

____

In R^3.

____

c) Do a series of experiments with vectors in R^2 using **uball**. Provide a verbal description of the unit ball for each value of p. (Choose the number of trials > 300.)

p	Description of Unit Ball.
3	______________________
4	______________________
5	______________________
10	______________________
20	______________________

d) In R^2 if $p \to \infty$, what geometric shape is the unit ball approaching?

____

e) Run **uball** in R^2 using the norm **inf**. Describe the unit ball.

____

f) Compare your descriptions in parts d) and e).

____

g) Do a series of experiments with vectors in R^3 using **uball**. Provide a verbal description of the unit ball for each value of p. (Choose the number of trials > 500.)

p	Description of Unit Ball.
3	______________________
4	______________________
5	______________________
10	______________________
20	______________________

h) In R^3 if p$\rightarrow \infty$, what geometric shape is the unit ball approaching?

____

i) Run **uball** in R^3 using the norm **inf**. Describe the unit ball.

____

j) Compare your descriptions in parts h) and i).

____

9. Run **uball** in R^2 using 1/2 for the 'norm choice' and observe the unit ball.[1] Repeat this process for $1/3, 1/4, 1/5, \ldots, 1/N$. If $N \rightarrow \infty$, what geometric shape is the unit ball approaching?

____

10. Run **uball** in R^3 using 1/2 for the 'norm choice' and observe the unit ball.[2] Repeat this process for $1/3, 1/4, 1/5, \ldots, 1/N$. If $N \rightarrow \infty$, what geometric shape is the unit ball approaching?

____

11. In MATLAB , let **x = rand(3,1)**, **t = sqrt(2)**, and

$$\boldsymbol{P} = \begin{bmatrix} 1/t & 1/t & 0 \\ 0 & 0 & 1 \\ 1/t & -1/t & 0 \end{bmatrix}$$

a) Compute $\boldsymbol{y} = \boldsymbol{Px}$. $\qquad \boldsymbol{y} =$ ______________

b) Compute $\| \boldsymbol{x} \|$ and $\| \boldsymbol{y} \|$.

$\| \boldsymbol{x} \| =$ ______________ $\qquad \| \boldsymbol{y} \| =$ ______________

[1] For values of p between 0 and 1, the expression given in Exercise 8 is not a norm. However, the geometry of the sets generated is interesting to study.

[2] *ibid*

c) Are $\| \boldsymbol{x} \|$ and $\| \boldsymbol{y} \|$ related? If so, how?

____

d) Experiment with other $\boldsymbol{x}$ using **rand(3,1)** to confirm or reject your answer to part c.

e) Summarize the relationship between the length of $\boldsymbol{x}$ and the length of $\boldsymbol{Px}$.

____

Section 9.3

Angles

Here we discuss a particular property of a real inner product space **V**, that is, a vector space **V** with real scalars and an inner product function that associates pairs of vectors from **V** with real numbers. R^n with the standard inner product is a real inner product space, but C^n with the standard inner product (see Exercise 3 in Section 9.1) is not. In any real inner product space **V** the *Cauchy-Schwarz* inequality holds. Hence for any pair of vectors $\boldsymbol{v}$ and $\boldsymbol{w}$ in **V**

$$(\boldsymbol{v}, \boldsymbol{w})^2 \leq (\boldsymbol{v}, \boldsymbol{v})(\boldsymbol{w}, \boldsymbol{w})$$

Assuming that the inner product is used to define a norm as $\|\boldsymbol{v}\| = \sqrt{(\boldsymbol{v}, \boldsymbol{v})}$ the Cauchy-Schwarz inequality implies that

$$|(\boldsymbol{v}, \boldsymbol{w})|^2 \leq \| \boldsymbol{v} \|^2 \| \boldsymbol{w} \|^2$$

or equivalently

$$|(\boldsymbol{v}, \boldsymbol{w})| \leq \| \boldsymbol{v} \| \| \boldsymbol{w} \|$$

If neither $\boldsymbol{v}$ nor $\boldsymbol{w}$ is the zero vector, then we have

$$\frac{|(\boldsymbol{v}, \boldsymbol{w})|}{\| \boldsymbol{v} \| \| \boldsymbol{w} \|} \leq 1$$

hence

$$-1 \leq \frac{(\boldsymbol{v}, \boldsymbol{w})}{\| \boldsymbol{v} \| \| \boldsymbol{w} \|} \leq 1$$

Thus we can use the quantity $\frac{(\boldsymbol{v},\boldsymbol{w})}{\|\boldsymbol{v}\|\|\boldsymbol{w}\|}$ to define the cosine of the angle between vectors $\boldsymbol{v}$ and $\boldsymbol{w}$. In a real inner product space $\mathbf{V}$ the cosine of the angle θ between vectors $\boldsymbol{v}$ and $\boldsymbol{w}$ is defined by

$$\cos\theta = \frac{(\boldsymbol{v},\boldsymbol{w})}{\parallel \boldsymbol{v} \parallel \parallel \boldsymbol{w} \parallel}$$

In MATLAB , for R^n with the standard inner product, $\cos\theta$ is computed by the command

dot(v,w)/(norm(v)*norm(w))

We compute the angle θ in radians by taking the arccosine of the previous expression. The arccosine command in MATLAB is **acos**.

<u>Example 1.</u> Let $\boldsymbol{u} = \begin{bmatrix} 3 \\ -1 \\ 0 \\ 2 \end{bmatrix}$ and $\boldsymbol{v} = \begin{bmatrix} 0 \\ 2 \\ 1 \\ -4 \end{bmatrix}$ be vectors in R^4. Enter these vectors as columns into MATLAB . Then the cosine of the angle between them is given by command

c = dot(u,v) / (norm(u)*norm(v))

which gives

```
c =

   -0.5832
```

and the angle in radians is obtained by command

angle = acos(c)

The display generated is

```
angle =

    2.1935
```

This output is in radians. To obtain an output in degrees type command **adeg = (acos(c)*180)/pi**. The display generated here is

```
adeg =

    125.7
```

The vectors $\boldsymbol{v}$ and $\boldsymbol{w}$ in a real inner product space are said to be *orthogonal* provided that the angle between them is $\pi/2$ radians. Computationally, $\boldsymbol{v}$ and $\boldsymbol{w}$ are orthogonal if and only if $(\boldsymbol{v}, \boldsymbol{w}) = \mathbf{0}$. Note that the vectors in Example 1 are not orthogonal.

Orthogonal vectors are important in a variety of situations and applications. The following example illustrates how to use a system of equations to determine vectors orthogonal to a given set of vectors.

Example 2. Let $\mathbf{V} = R^3$ with the standard inner product. Let $\mathbf{S}=\{\boldsymbol{v}_1, \boldsymbol{v}_2\} = \left\{ \begin{bmatrix} 1 \\ 2 \\ 1 \end{bmatrix}, \begin{bmatrix} 2 \\ 1 \\ 0 \end{bmatrix} \right\}$.

Find a vector orthogonal to each vector in $\mathbf{S}$.

Let $\mathbf{c} = \begin{bmatrix} c_1 \\ c_2 \\ c_3 \end{bmatrix}$ and require $\mathbf{c}$ to be chosen so that $(\boldsymbol{v}_1, \mathbf{c}) = 0$ and $(\boldsymbol{v}_2, \mathbf{c}) = 0$. This leads to the homogeneous linear system

$$\begin{bmatrix} 1 & 2 & 1 \\ 2 & 1 & 0 \end{bmatrix} \begin{bmatrix} c_1 \\ c_2 \\ c_3 \end{bmatrix} = \begin{bmatrix} 0 \\ 0 \end{bmatrix}.$$

Hence $\mathbf{c}$ is in the null space of matrix $\boldsymbol{A} = \begin{bmatrix} 1 & 2 & 1 \\ 2 & 1 & 0 \end{bmatrix}$. From MATLAB , command **rref(A)** gives $\begin{bmatrix} 1 & 0 & -1/3 \\ 0 & 1 & 2/3 \end{bmatrix}$ which implies that the general solution to $\boldsymbol{Ac} = \mathbf{0}$ is $\begin{bmatrix} r/3 \\ -2r/3 \\ r \end{bmatrix}$. Hence one vector orthogonal to each vector in $\mathbf{S}$ is $\boldsymbol{c} = \begin{bmatrix} 1 \\ -2 \\ 3 \end{bmatrix}$ (set $r = 3$). Note that $\mathbf{c}$ is also orthogonal to span $\mathbf{S}$ because $\mathbf{c}$ is orthogonal to both $\boldsymbol{v}_1$ and $\boldsymbol{v}_2$. Note also that the set $\{\boldsymbol{v}_1, \boldsymbol{v}_2, \mathbf{c}\}$ forms a basis for R^3.

Exercises 9.3

1. Find the angle (in radians) between each of the following pairs of vectors in R^n using the standard inner product. Record your result in the space provided.

a) $v = \begin{bmatrix} 1 \\ 2 \end{bmatrix}$, $w = \begin{bmatrix} 1 \\ 3 \end{bmatrix}$. ______

b) $v = \begin{bmatrix} 1 \\ 1 \end{bmatrix}$, $w = \begin{bmatrix} 1 \\ -1 \end{bmatrix}$. ______

c) $v = \begin{bmatrix} 2 \\ 1 \\ 3 \end{bmatrix}$, $w = \begin{bmatrix} 1 \\ -2 \\ 1 \end{bmatrix}$. ______

d) $v = \begin{bmatrix} 1 \\ 2 \\ 0 \\ 3 \end{bmatrix}$, $w = \begin{bmatrix} 2 \\ -1 \\ 4 \\ 0 \end{bmatrix}$. ______

e) $v = \begin{bmatrix} 4 \\ -4 \\ 2 \\ -2 \end{bmatrix}$, $w = \begin{bmatrix} 1 \\ 2 \\ -3 \\ -1 \end{bmatrix}$. ______

f) $v = \begin{bmatrix} 3 \\ 1 \\ 2 \\ 0 \end{bmatrix}$, $w = \begin{bmatrix} -1 \\ -1 \\ 2 \\ 1 \end{bmatrix}$. ______

2. Let $\mathbf{V} = R^2$ with inner product $(\boldsymbol{v}, \boldsymbol{u}) = \boldsymbol{v}^T \boldsymbol{A} \boldsymbol{u}$ where $\boldsymbol{A} = \begin{bmatrix} 3 & 1 \\ 1 & 3 \end{bmatrix}$. (It can be shown that this definition satisfies all the axioms of an inner product.) Compute the angle between the following pairs of vectors in this inner product space and record your results in the space provided. (In MATLAB the inner product defined here can be computed as **dot(v,A*u)**.)

a) $v = \begin{bmatrix} 1 \\ 2 \end{bmatrix}$, $u = \begin{bmatrix} 1 \\ 3 \end{bmatrix}$. ______

b) $v = \begin{bmatrix} 1 \\ 1 \end{bmatrix}$, $u = \begin{bmatrix} 1 \\ -1 \end{bmatrix}$. ______

c) $v = \begin{bmatrix} 1 \\ 0 \end{bmatrix}$, $u = \begin{bmatrix} 3 \\ 2 \end{bmatrix}$. ______

3. Let **V**= R^4 with the standard inner product and let

$$\boldsymbol{v_1} = \begin{bmatrix} 3 \\ 1 \\ 2 \\ 0 \end{bmatrix}, \quad \boldsymbol{v_2} = \begin{bmatrix} 1 \\ -1 \\ 0 \\ 1 \end{bmatrix}, \quad \boldsymbol{v_3} = \begin{bmatrix} 2 \\ 0 \\ 0 \\ 1 \end{bmatrix}.$$

a) Let $\mathbf{T} = \{\boldsymbol{v_1}, \boldsymbol{v_2}, \boldsymbol{v_3}\}$. Find a vector $\boldsymbol{w}$ orthogonal to each vector in **T**. Describe your procedure and record the vectors you find.

b) Let $\mathbf{S} = \{\boldsymbol{v_1}, \boldsymbol{v_2}\}$. Find two linearly independent vectors $\boldsymbol{w}$ and $\boldsymbol{u}$ that are orthogonal to each vector in **S**. Describe your procedure and record the vectors you find.

LAB 10

Orthogonal Sets

Topics: orthogonal sets; orthonormal sets; coordinates of a vector relative to an orthonormal basis; projections; construction of an orthonormal basis using the Gram-Schmidt process; command **gschmidt**.

Introduction

Let $\mathbf{V}$ be an inner product space with the inner product of a pair of vectors $\boldsymbol{u}$ and $\boldsymbol{v}$ in $\mathbf{V}$ denoted by $(\boldsymbol{u}, \boldsymbol{v})$. From Section 9.3 we have that $\boldsymbol{u}$ and $\boldsymbol{v}$ are orthogonal provided $(\boldsymbol{u}, \boldsymbol{v}) = 0$. Here we consider sets of orthogonal vectors and investigate bases which are orthogonal sets.

Let $\mathbf{S} = \{\boldsymbol{v}_1, \boldsymbol{v}_2, \ldots, \boldsymbol{v}_k\}$ be a set of vectors in $\mathbf{V}$.

- $\mathbf{S}$ is called an *orthogonal set* provided $(\boldsymbol{v}_i, \boldsymbol{v}_j) = 0$ for $i \neq j$. (We say that the vectors in $\mathbf{S}$ are mutually orthogonal.)

- If $\mathbf{S}$ is an orthogonal set of nonzero vectors, then $\mathbf{S}$ is linearly independent.

- If $\mathbf{S}$ is an orthogonal set of nonzero vectors, then

$$\mathbf{T} = \left\{ \frac{\boldsymbol{v}_1}{\| \boldsymbol{v}_1 \|}, \frac{\boldsymbol{v}_2}{\| \boldsymbol{v}_2 \|}, \ldots, \frac{\boldsymbol{v}_k}{\| \boldsymbol{v}_k \|} \right\}$$

  is an orthogonal set in which each vector has length one, where $\| \boldsymbol{v}_j \| = \sqrt{(\boldsymbol{v}_j, \boldsymbol{v}_j)}$. The operation of dividing a nonzero vector by its length is referred to as *normalizing* the vector.

- A set of vectors which is orthogonal and in which each vector has length one is called an *orthonormal set.* (The set $\mathbf{T}$ above is an orthonormal set.)

Section 10.1 discusses matrices whose columns form an orthonormal basis. Such matrices are called orthogonal and play important roles in a variety of topics.

Section 10.2 develops the projection of one vector onto another and the projection of a vector onto a subspace. We use both a geometric and algebraic approach and show the important role of orthogonal bases. Projections provide an important way to obtain approximations.

Section 10.3 develops an algorithm, the Gram-Schmidt process, for producing an orthonormal basis from an existing basis for a subspace. In effect this shows that the projection techniques from Section 10.2 can always be applied.

Section 10.1

Orthonormal Bases

Let $\mathbf{S} = \{\boldsymbol{u}_1, \boldsymbol{u}_2, \ldots, \boldsymbol{u}_n\}$ be an orthonormal basis for an inner product space $\mathbf{V}$. Then for any vector $\boldsymbol{v}$ in $\mathbf{V}$ it is easy to compute the coordinates of $\boldsymbol{v}$ relative to $\mathbf{S}$ using the inner product. Suppose that we want to find $c_i, i = 1, 2, \ldots, n$ such that

$$\boldsymbol{v} = c_1\boldsymbol{u}_1 + c_2\boldsymbol{u}_2 + \cdots + c_n\boldsymbol{u}_n$$

Then using the property that the vectors in **S** form an orthonormal set of vectors, upon taking the inner product of each side of the previous expression with $\boldsymbol{u_k}$ we have that

$$c_k = (\boldsymbol{v}, \boldsymbol{u_k}), \qquad \text{for } k = 1, 2, ..., n.$$

Now let $\boldsymbol{T}$ be an $n \times n$ matrix whose jth column is denoted by $\boldsymbol{w_j}$. It is instructive to view $\boldsymbol{T}$ as

$$\boldsymbol{T} = \begin{bmatrix} \boldsymbol{w}_1 & \boldsymbol{w}_2 & \cdots & \boldsymbol{w}_n \end{bmatrix}$$

The definition of matrix multiplication enables the entries of the matrix $\boldsymbol{T}'*\boldsymbol{T}$ to be written in terms of the inner products $(\boldsymbol{w_i}, \boldsymbol{w_j})$.

$$\boldsymbol{T}' * \boldsymbol{T} = \begin{bmatrix} \boldsymbol{w}_1' \\ \boldsymbol{w}_2' \\ \vdots \\ \boldsymbol{w}_n' \end{bmatrix} \begin{bmatrix} \boldsymbol{w}_1 & \boldsymbol{w}_2 & \cdots & \boldsymbol{w}_n \end{bmatrix} = \begin{bmatrix} (\boldsymbol{w}_1, \boldsymbol{w}_1) & (\boldsymbol{w}_1, \boldsymbol{w}_2) & \cdots & (\boldsymbol{w}_1, \boldsymbol{w}_n) \\ (\boldsymbol{w}_2, \boldsymbol{w}_1) & (\boldsymbol{w}_2, \boldsymbol{w}_2) & \cdots & (\boldsymbol{w}_2, \boldsymbol{w}_n) \\ \vdots & \vdots & \vdots & \\ (\boldsymbol{w}_n, \boldsymbol{w}_1) & (\boldsymbol{w}_n, \boldsymbol{w}_2) & \cdots & (\boldsymbol{w}_n, \boldsymbol{w}_n) \end{bmatrix}$$

Two results follow immediately from this form.

- The columns of $\boldsymbol{T}$ form an orthogonal set if and only if $\boldsymbol{T}'*\boldsymbol{T}$ is a diagonal matrix.

- The columns of $\boldsymbol{T}$ form an orthonormal set if and only if $\boldsymbol{T}'*\boldsymbol{T} = \boldsymbol{I_n}$.

  The preceding statement is equivalent to the following:

  The columns of $\boldsymbol{T}$ form an orthonormal set if and only if $\boldsymbol{T}^{-1} = \boldsymbol{T}'$.

Square matrices with orthonormal columns are important in a number of areas and arise later in this chapter. We emphasize this with the following terminology.

> A square matrix $\boldsymbol{P}$ with real entries is called **orthogonal** provided $\boldsymbol{P}' = \boldsymbol{P}^{-1}$.

Example 1. Let $\mathbf{S} = \{\boldsymbol{v}_1, \boldsymbol{v}_2, \boldsymbol{v}_3\} = \left\{ \begin{bmatrix} 1 \\ 1 \\ 0 \end{bmatrix}, \begin{bmatrix} 1 \\ -1 \\ 2 \end{bmatrix}, \begin{bmatrix} -1 \\ 1 \\ 1 \end{bmatrix} \right\}$. Enter these vectors into MATLAB as **v1, v2, v3** respectively. To verify that $\boldsymbol{S}$ is an orthogonal set you can use the definition and check that in MATLAB **dot(v1, v2), dot(v2, v3),** and **dot(v1, v3)** are all equal to zero, or in MATLAB form the matrix

C – [v1 v2 v3]

and perform the multiplication $C' * C$ to check that the product is equal to the diagonal matrix

$$\begin{bmatrix} 2 & 0 & 0 \\ 0 & 6 & 0 \\ 0 & 0 & 3 \end{bmatrix}$$

Since $C' * C$ is not equal to the identity matrix, the set **S** is not orthonormal. To normalize set **S**, enter MATLAB commands

```
t1 = (1/norm(v1))*v1
t2 = (1/norm(v2))*v2
t3 = (1/norm(v3))*v3
```

The set **T** = **{t1, t2, t3}** is orthonormal. To verify this, construct the matrix A as

```
A = [t1 t2 t3]
```

in MATLAB and then compute $A' * A$. You will see the 3×3 identity matrix displayed.

Let $v = [15 \ -7 \ 7]'$. The coordinates of v relative to basis S are $[v]_S = [4 \ 6 \ -5]'$, which can be obtained from **rref([C v])** (See Section 7.2 for details on coordinates of a vector relative to a basis.) or from $C\backslash v$. It is possible to obtain $[v]_T$ in a similar manner. However, since T is an orthogonal basis, $[v]_T$ can be computed directly using $[v]_T = [(v, t1) \ (v, t2) \ (v, t3)]'$.

Exercises 10.1

1. Let $\mathbf{V} = R^3$ with the standard inner product and let

$$\mathbf{S} = \{u_1, u_2, u_3\} = \left\{ \begin{bmatrix} 1/\sqrt{3} \\ 1/\sqrt{3} \\ 1/\sqrt{3} \end{bmatrix}, \begin{bmatrix} -2/\sqrt{6} \\ 1/\sqrt{6} \\ 1/\sqrt{6} \end{bmatrix}, \begin{bmatrix} 0 \\ -1/\sqrt{2} \\ 1/\sqrt{2} \end{bmatrix} \right\}.$$

In MATLAB u_1 can be entered by typing **t = 1/sqrt(3); u1 = [t; t; t]** and similarly for **u2** and **u3**.

a) Using MATLAB show that **S** is an orthonormal basis for **V**. Write a brief statement indicating your approach.

____

____

b) For $v = \begin{bmatrix} 1 \\ 2 \\ 3 \end{bmatrix}$, find $[v]_S$ ____________________

c) For $v = \begin{bmatrix} -1 \\ 0 \\ 4 \end{bmatrix}$, find $[v]_S$ ____________________

2. Let $\mathbf{V} = R^2$ with the standard inner product and let

$$\mathbf{S} = \{\boldsymbol{u}_1, \boldsymbol{u}_2\} = \left\{ \begin{bmatrix} 1 \\ 1 \end{bmatrix}, \begin{bmatrix} -1 \\ 1 \end{bmatrix} \right\}$$

be a basis for $\mathbf{V}$.

a) Using MATLAB show that $\mathbf{S}$ is an orthogonal set. (State how to do this on the line below.)

____

b) Convert the $\mathbf{S}$-basis to an orthonormal basis. Call the new basis $\mathbf{T}$ and display its vectors below.

c) Let $\boldsymbol{w} = \begin{bmatrix} 2 \\ -3 \end{bmatrix}$. Find the coordinate vector of $\boldsymbol{w}$ relative to the $\mathbf{T}$-basis.

$[\boldsymbol{w}]_T =$ ____________________.

d) What is the coordinate vector of $\boldsymbol{w}$ relative to the $\mathbf{S}$-basis?

$[\boldsymbol{w}]_S =$ ____________________.

e) How are the coordinate vectors in c) and d) related?

____

3. In MATLAB use command **x = rand** and then form the matrix

$$A = \begin{bmatrix} \cos(x) & -\sin(x) \\ \sin(x) & \cos(x) \end{bmatrix}$$

a) Compute $A' * A$. ____________________

b) The set of columns of A forms an ____________________ set.

c) Matrix A is an ____________________ matrix.

4. Is the set of columns of the matrix generated by the MATLAB command **H = hilb(5)** an orthogonal set? Explain.

____

5. Let $v_1 = \begin{bmatrix} 1/\sqrt{3} \\ 1/\sqrt{3} \\ 1/\sqrt{3} \end{bmatrix}$ and $v_2 = \begin{bmatrix} -1/\sqrt{2} \\ 0 \\ 1/\sqrt{2} \end{bmatrix}$. Find a vector v_3 so that the set $S = \{v_1, v_2, v_3\}$ is orthonormal.

6. Let $A = \begin{bmatrix} 1 & -1 & p \\ 1 & 1 & q \\ 0 & 1 & r \end{bmatrix}$.

a) Determine values p, q, and r so that $A'A$ is a diagonal matrix. How many such values are there?

b) If $r = 1$, then find an orthogonal matrix whose columns are scalar multiples of A.

7. Let P be an $n \times n$ orthogonal matrix and x and y be vectors in R^n.

a) Show that $\| Px \| = \| x \|$.

b) Show that the angle between Px and Py is the same as the angle between x and y.

Section 10.2

Projections

Here we investigate the concept of the (orthogonal) projection of one vector onto another from both a geometrical and a computational standpoint. We use an intuitive development based on trigonometric tools in R^2 and extend the procedures to R^3 and beyond. We conclude our discussion with the projection of a vector onto a plane to set the stage for projections onto subspaces in the next section. Both the computational power and the graphics capability of MATLAB will be used to provide a foundation for the important notion of a projection.

In R^2:

A geometric point of view.

The projection of a vector $\boldsymbol{u}$ onto a vector $\boldsymbol{w}$ is obtained by dropping a perpendicular from the tip of $\boldsymbol{u}$ onto $\boldsymbol{w}$. See Figure 1. (If needed we extend $\boldsymbol{w}$. See Figure 2.) Note that the

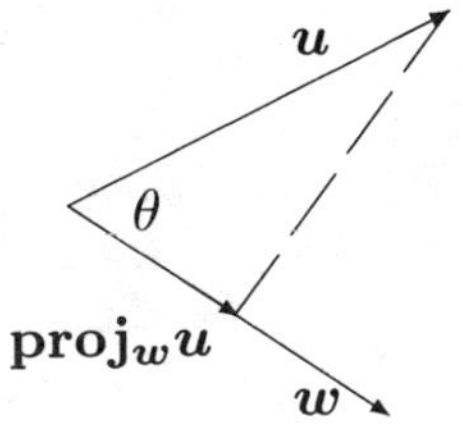

Figure 1

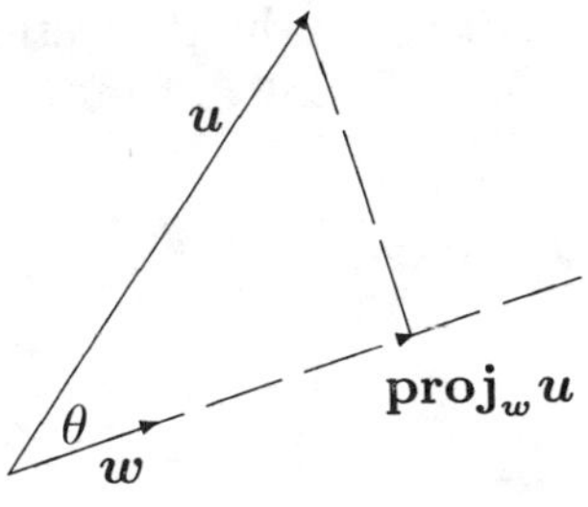

Figure 2

projection in these pictures is in the same direction [1] as $\boldsymbol{w}$ and by using trigonometry the length of the projection is $\cos\theta \parallel \boldsymbol{u} \parallel$. We adopt the following notation. The projection of vector $\boldsymbol{u}$ onto vector $\boldsymbol{w}$ is a vector denoted by

$$\mathbf{proj}_{\boldsymbol{w}}\ \boldsymbol{u}$$

An algebraic point of view.

We know the length of the projection is $\cos\theta \parallel \boldsymbol{u} \parallel$ and that the projection is a vector in the same direction as $\boldsymbol{w}$. Thus algebraically we can express the projection as its length times a unit vector which has the same direction. Hence we have the expression

$$\mathbf{proj}_{\boldsymbol{w}}\ \boldsymbol{u} = \cos\theta \parallel \boldsymbol{u} \parallel \frac{\boldsymbol{w}}{\parallel \boldsymbol{w} \parallel}$$

[1] Assume that the angle between $\boldsymbol{u}$ and $\boldsymbol{w}$ is greater than $\pi/2$ radians. Draw the figure and the projection in this case. Explain why in general we say that the projection of vector $\boldsymbol{u}$ onto vector $\boldsymbol{w}$ is a vector parallel to $\boldsymbol{w}$.

Next we use the formula for the cosine of the angle between two vectors (see the Cauchy-Schwarz inequality)

$$\cos\theta = \frac{(\boldsymbol{u}, \boldsymbol{w})}{\| \boldsymbol{u} \| \| \boldsymbol{w} \|}$$

to obtain another formula for the projection. Substituting for $\cos\theta$ we have

$$\mathbf{proj}_{\boldsymbol{w}}\ \boldsymbol{u} = \cos\theta \| \boldsymbol{u} \| \frac{\boldsymbol{w}}{\| \boldsymbol{w} \|} = \frac{(\boldsymbol{u}, \boldsymbol{w})}{\| \boldsymbol{u} \| \| \boldsymbol{w} \|} \| \boldsymbol{u} \| \frac{\boldsymbol{w}}{\| \boldsymbol{w} \|}$$

Simplifying we have

$$\mathbf{proj}_{\boldsymbol{w}}\ \boldsymbol{u} = \frac{(\boldsymbol{u}, \boldsymbol{w})}{\| \boldsymbol{w} \|^2} \boldsymbol{w} \tag{10.1}$$

Example 1. Let $\boldsymbol{u} = \begin{bmatrix} 3 \\ 2 \end{bmatrix}$ and $\boldsymbol{w} = \begin{bmatrix} 4 \\ -1 \end{bmatrix}$. Find $\mathbf{proj}_{\boldsymbol{w}}\ \boldsymbol{u}$.

Using Equation (10.1) for the projection we compute the inner product of vectors $\boldsymbol{u}$ and $\boldsymbol{w}$ and the length of $\boldsymbol{w}$. In MATLAB , command

dot(u,w)

displays

```
ans =

    10
```

and command

norm(w)

displays

```
ans =

    4.1231
```

It follows that $\|\boldsymbol{w}\|^2 = 17$. (In MATLAB use command **ans**^2.) Hence from Equation 10.1,

$$\mathbf{proj}_{\boldsymbol{w}}\ \boldsymbol{u} = \frac{10}{17}\boldsymbol{w} = \frac{10}{17}\begin{bmatrix} 4 \\ -1 \end{bmatrix}.$$

To see the projection of $\boldsymbol{u}$ onto $\boldsymbol{w}$ use MATLAB command

project(u,w)

The graphics display in two dimensions in routine **project** shows a line drawing similar to Figure 1 with $\mathbf{proj}_{\boldsymbol{w}}\ \boldsymbol{u}$ labeled by $\boldsymbol{P}$. Note that the line segment perpendicular to $\boldsymbol{w}$ can be written in terms of $\boldsymbol{u}$ and the projection $\boldsymbol{P}$ as $-\boldsymbol{P}+\boldsymbol{u}$. (For more information on routine **project** use **help**.)

A vector space point of view.

In R^2 $\mathbf{span}\{\boldsymbol{w}\}$ is a subspace which geometrically is a line through the origin. The projection of $\boldsymbol{u}$ onto $\boldsymbol{w}$ is a vector in $\mathbf{span}\{\boldsymbol{w}\}$. Since we measure distances as perpendicular distance we see that the vector $\mathbf{proj}_{\boldsymbol{w}}\ \boldsymbol{u}$ is 'closest' to $\boldsymbol{u}$ since vector $\boldsymbol{s}$ (See Figure 3.) is orthogonal to $\mathbf{span}\{\boldsymbol{w}\}$. Hence the projection of $\boldsymbol{u}$ onto $\boldsymbol{w}$ represents the member of the subspace $\mathbf{span}\{\boldsymbol{w}\}$ that is closest to vector $\boldsymbol{u}$. It also follows from standard vector considerations that we can obtain

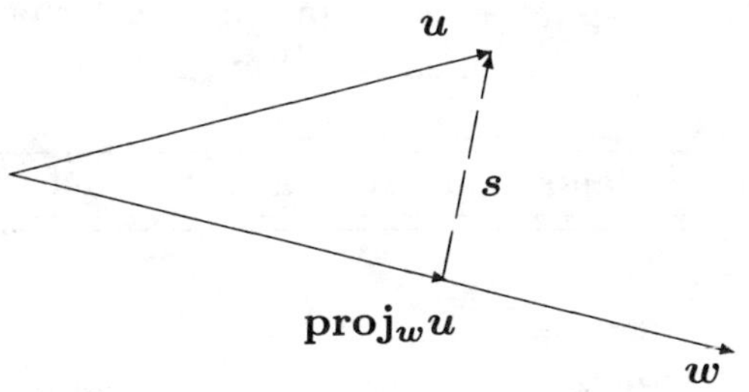

Figure 3

a formula for vector $\boldsymbol{s}$. Vector $\boldsymbol{s}$ is a vector orthogonal to every vector in $\mathbf{span}\{\boldsymbol{w}\}$. We have

$$\boldsymbol{u} \;=\; \mathbf{proj}_{\boldsymbol{w}}\ \boldsymbol{u} \;+\; \boldsymbol{s}$$

thus

$$\boldsymbol{s} = \boldsymbol{u} - \mathbf{proj}_{\boldsymbol{w}}\ \boldsymbol{u} \tag{10.2}$$

Example 2. Using the vectors $\boldsymbol{u}$ and $\boldsymbol{w}$ in Example 1, find a vector that is orthogonal to $\mathbf{span}\{\boldsymbol{w}\}$. Using Equation(10.2) we have that a vector $\boldsymbol{s}$ orthogonal to $\mathbf{span}\{\boldsymbol{w}\}$ is

$$\boldsymbol{s} = \boldsymbol{u} - \mathbf{proj}_{\boldsymbol{w}}\ \boldsymbol{u} = \begin{bmatrix} 3 \\ 2 \end{bmatrix} - \frac{10}{17}\begin{bmatrix} 4 \\ -1 \end{bmatrix} = \begin{bmatrix} 11/17 \\ 44/17 \end{bmatrix}$$

To check, compute the inner product of $\boldsymbol{s}$ with any vector in $\mathbf{span}\{\boldsymbol{w}\}$. Here $\mathbf{span}\{\boldsymbol{w}\}$ is any multiple of vector $\boldsymbol{w}$, so let $k\boldsymbol{w}$ represent any vector in $\mathbf{span}\{\boldsymbol{w}\}$.

$$(\boldsymbol{s},k\boldsymbol{w}) = \frac{11}{17}(4k) + \frac{44}{17}(-k) = 0$$

If you used routine **project** in Example 1 (and have not exited MATLAB or done other graphics) type command **figure(gcf)** to redisplay the line drawing of the projection process. Note that $\boldsymbol{s}$, the line segment perpendicular to $\boldsymbol{w}$ in Figure 3, can be used so that $\textbf{span}\{\boldsymbol{w}, \boldsymbol{s}\} = \textbf{span}\{\boldsymbol{u}, \boldsymbol{w}\} = R^2$. Hence set $\{\boldsymbol{w}, \boldsymbol{s}\}$ is an orthogonal basis for R^2 while set $\{\boldsymbol{u}, \boldsymbol{w}\}$ is a basis, but not an orthogonal set. Press ENTER to return to the command screen, then type the following command.

gtext('s')

Move the cross-hair using the arrow keys to label the vector $\boldsymbol{s}$ which is orthogonal to $\boldsymbol{w}$. Press ENTER to return to the command screen. (Use **help** for more information on **gtext**.)

Summary: Projections give us a way to compute a vector in $\textbf{span}\{\boldsymbol{w}\}$ that is closest to $\boldsymbol{u}$ and a way to find a vector that is orthogonal to every member of $\textbf{span}\{\boldsymbol{w}\}$.

These ideas can be extended to R^n and to any real inner product space!

In R^3, the projection of vector $\boldsymbol{u} = \begin{bmatrix} u_1 \\ u_2 \\ u_3 \end{bmatrix}$ onto vector $\boldsymbol{w} = \begin{bmatrix} w_1 \\ w_2 \\ w_3 \end{bmatrix}$ is obtained by employing the identical formula as in Equation (10.1). We use MATLAB and its graphics to illustrate the process.

Example 3. Let $\boldsymbol{u} = \begin{bmatrix} 4 \\ 5 \\ 6 \end{bmatrix}$ and $\boldsymbol{w} = \begin{bmatrix} 7 \\ 7 \\ -3 \end{bmatrix}$. To determine $\textbf{proj}_{\boldsymbol{w}}\ \boldsymbol{u}$ use MATLAB command

p = (dot(u,w)/norm(w)^2)*w

To see the components of projection $\boldsymbol{p}$ in rational form type **rational(p)**. To display a line drawing of the process use command

project(u,w)

Use the command **gtext('s')** as described in Example 2 to label vector $\boldsymbol{s}$ on the displayed figure.

To extend the ideas on projections to subspaces of R^3 we proceed as follows. Let $\boldsymbol{u}$ be a vector in R^3 which we want to project onto a plane. A plane is a subspace of dimension 2 in

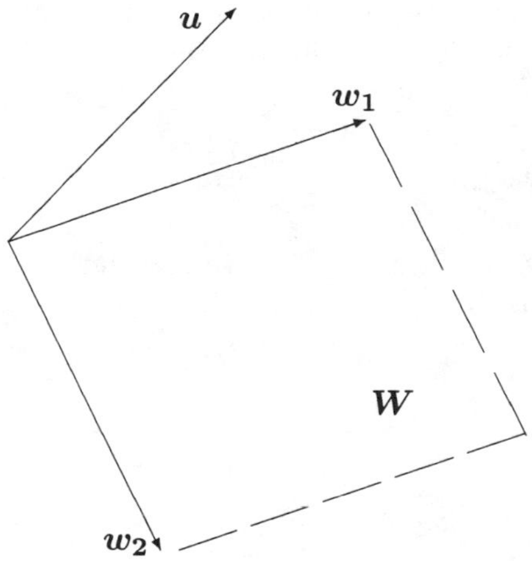

Figure 4

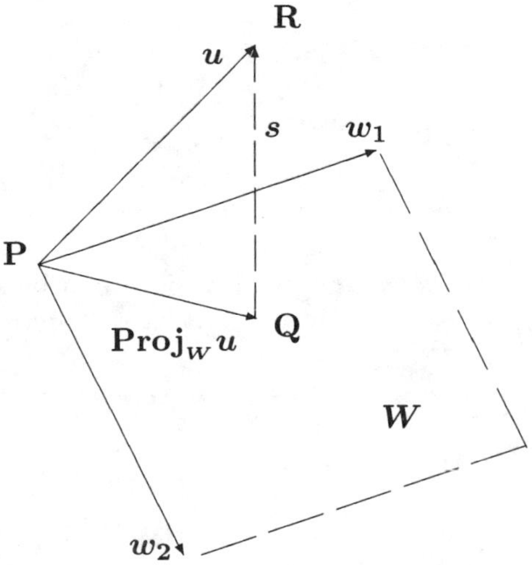

Figure 5

R^3 and is completely defined if we specify a basis. Let $\boldsymbol{w_1}$ and $\boldsymbol{w_2}$ be linearly independent **orthogonal vectors** in R^3 so that $\mathbf{W}=$ **span**$\{\boldsymbol{w_1}, \boldsymbol{w_2}\}$ and $\boldsymbol{u}$ is not in $\mathbf{W}$. The situation is illustrated in Figure 4. The projection of $\boldsymbol{u}$ onto subspace $\mathbf{W}$ is obtained by dropping a perpendicular from the tip of $\boldsymbol{u}$, denoted by R, until we intersect the plane $\mathbf{W}$ at point Q. In Figure 5 we have indicated the projection by connecting point P with Q. The vector from Q to R is labeled $\boldsymbol{s}$ and is orthogonal to the plane $\mathbf{W}$. Since $\mathbf{W}$ is a subspace, $\boldsymbol{s}$ is orthogonal to $\mathbf{W}$ if and only if it is orthogonal to every vector in the plane $\mathbf{W}$. But $\mathbf{W} =$ **span**$\{\boldsymbol{w_1}, \boldsymbol{w_2}\}$, hence $\boldsymbol{s}$ is orthogonal to every linear combination of vectors $\boldsymbol{w_1}$ and $\boldsymbol{w_2}$. Figure 5 provides us

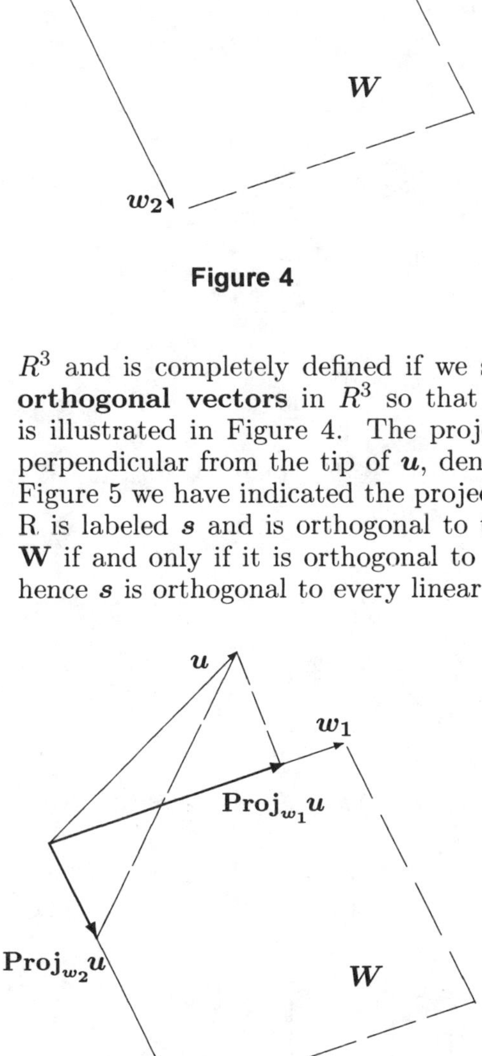

Figure 6

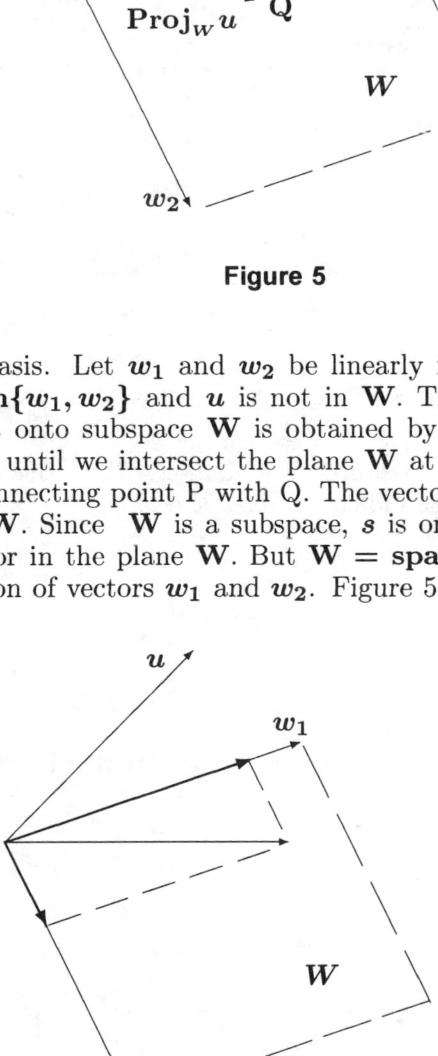

Figure 7

with a picture of the projection $\mathbf{proj_W}\ \boldsymbol{u}$, but to work with the projection we need an algebraic expression for $\mathbf{proj_W}\ \boldsymbol{u}$. Since a basis gives us the minimal amount of information about a subspace it follows that we should consider projecting $\boldsymbol{u}$ onto the basis vectors for $\mathbf{W}$. This is shown in Figure 6. From the previous work in R^2, and the fact that things generalize in a natural fashion, we have

$$\mathbf{proj}_{\boldsymbol{w}_1}\ \boldsymbol{u} = \frac{(\boldsymbol{u}, \boldsymbol{w}_1)}{\| \boldsymbol{w}_1 \|^2}\boldsymbol{w}_1 \qquad \text{and} \qquad \mathbf{proj}_{\boldsymbol{w}_2}\ \boldsymbol{u} = \frac{(\boldsymbol{u}, \boldsymbol{w}_2)}{\| \boldsymbol{w}_2 \|^2}\boldsymbol{w}_2$$

Since we have an orthogonal basis it can be shown that $\mathbf{proj}_{\mathbf{W}}\ \boldsymbol{u}$ can be written as the sum of the projections of $\boldsymbol{u}$ onto the basis vectors for $\mathbf{W}$. See Figure 7. We have

$$\mathbf{proj}_{\mathbf{W}}\ \boldsymbol{u} = \mathbf{proj}_{\boldsymbol{w}_1}\ \boldsymbol{u} + \mathbf{proj}_{\boldsymbol{w}_2}\ \boldsymbol{u} \tag{10.3}$$

or equivalently

$$\mathbf{proj}_{\mathbf{W}}\ \boldsymbol{u} = \frac{(\boldsymbol{u}, \boldsymbol{w}_1)}{\| \boldsymbol{w}_1 \|^2}\boldsymbol{w}_1 + \frac{(\boldsymbol{u}, \boldsymbol{w}_2)}{\| \boldsymbol{w}_2 \|^2}\boldsymbol{w}_2$$

Warning: This result is true only when $\boldsymbol{w}_1$ and $\boldsymbol{w}_2$ are orthogonal.

Example 4. Let $\mathbf{W} = \mathbf{span}\{\boldsymbol{w}_1, \boldsymbol{w}_2\} = \mathbf{span}\left\{\begin{bmatrix}1\\2\\1\end{bmatrix}, \begin{bmatrix}-2.5\\1\\.5\end{bmatrix}\right\}$ and $\boldsymbol{u} = \begin{bmatrix}-2\\2\\3\end{bmatrix}$. The set $\{\boldsymbol{w}_1, \boldsymbol{w}_2\}$ is an orthogonal set. Find $\mathbf{proj}_{\mathbf{W}}\ \boldsymbol{u}$ and a vector orthogonal to subspace $\mathbf{W}$. Proceeding as indicated above we have

$$\mathbf{proj}_{\boldsymbol{w}_1}\ \boldsymbol{u} = \frac{(\boldsymbol{u}, \boldsymbol{w}_1)}{\| \boldsymbol{w}_1 \|^2}\boldsymbol{w}_1 = \frac{5}{6}\boldsymbol{w}_1$$

$$\mathbf{proj}_{\boldsymbol{w}_2}\ \boldsymbol{u} = \frac{(\boldsymbol{u}, \boldsymbol{w}_2)}{\| \boldsymbol{w}_2 \|^2}\boldsymbol{w}_2 = \frac{17}{15}\boldsymbol{w}_2$$

and it follows that from Equation (10.3) that

$$\mathbf{proj}_{\mathbf{W}}\ \boldsymbol{u} = \frac{5}{6}\boldsymbol{w}_1 + \frac{17}{15}\boldsymbol{w}_2 = \begin{bmatrix}-2\\2.8\\1.4\end{bmatrix}$$

From Figure 5 we see that vector $\boldsymbol{s}$ is orthogonal to $\mathbf{W}$ and that

$$\boldsymbol{s} = \boldsymbol{u} - \mathbf{proj}_{\mathbf{W}}\ \boldsymbol{u} = \begin{bmatrix}-2\\2\\3\end{bmatrix} - \begin{bmatrix}-2\\2.8\\1.4\end{bmatrix} = \begin{bmatrix}0\\-0.8\\1.6\end{bmatrix}$$

In Example 4 the projection of $\boldsymbol{u}$ was onto the plane $\mathbf{W} = \mathbf{span}\{\boldsymbol{w}_1, \boldsymbol{w}_2\}$. For illustrative purposes we can think of plane $\mathbf{W}$ as the **xy**-plane. For in fact we can rotate the coordinate system so that $\mathbf{W}$ is rotated into the **xy**-plane. A graphical display of the projection of a vector $\boldsymbol{u}$ onto the **xy**-plane is available in MATLAB . Enter vector $\boldsymbol{u}$ into MATLAB , then type

projxy(u)

Follow the screen directions. Try this with vector $\boldsymbol{u}$ in Example 4.

The key to projections is to use an orthogonal basis.

In fact things are even simpler if we use an orthonormal basis, for then the denominators in Equation (10.3) are 1.

Note: For an orthonormal set of columns $\boldsymbol{w_1}, \boldsymbol{w_2}, \ldots, \boldsymbol{w_k}$ with $\boldsymbol{W} = \text{span}\{\boldsymbol{w_1}, \boldsymbol{w_2}, \ldots, \boldsymbol{w_k}\}$, the expression in 10.3 can be computed as follows using matrix $\boldsymbol{M} = [\boldsymbol{w_1}, \boldsymbol{w_2}, \ldots, \boldsymbol{w_k}]$

$$\mathbf{proj}_{\boldsymbol{W}}\boldsymbol{u} = \boldsymbol{M}(\boldsymbol{M}^T\boldsymbol{u}) = \boldsymbol{M}\begin{bmatrix} \boldsymbol{w}_1^T\boldsymbol{u} \\ \boldsymbol{w}_2^T\boldsymbol{u} \\ \vdots \\ \boldsymbol{w}_k^T\boldsymbol{u} \end{bmatrix} = (\boldsymbol{w}_1^T\boldsymbol{u})\boldsymbol{w}_1 + (\boldsymbol{w}_2^T\boldsymbol{u})\boldsymbol{w}_2 + \cdots + (\boldsymbol{w}_k^T\boldsymbol{u})\boldsymbol{w}_k\,. \tag{10.4}$$

Exercises 10.2

1. Let $\boldsymbol{w} = \begin{bmatrix} 3 \\ 4 \end{bmatrix}$.

a) Find the projection $\boldsymbol{p}$ of $\boldsymbol{u} = \begin{bmatrix} 1 \\ 2 \end{bmatrix}$ onto $\boldsymbol{w}$. $\boldsymbol{p} =$ ____________

b) Use command **project(u,w)** to determine the quadrant in which $\boldsymbol{p}$ is located.

Record the quadrant here. ____________

c) Find a scalar k for which the vector $k\boldsymbol{p}$ has a norm that is equal to one.

$k =$ ____________

d) Find a vector $\boldsymbol{s}$ that is orthogonal to **span**$\{\boldsymbol{w}\}$. $\boldsymbol{s} =$ ____________

2. Let $\boldsymbol{w} = \begin{bmatrix} 3 \\ 4 \end{bmatrix}$.

a) Find the projection $\boldsymbol{p}$ of $\boldsymbol{u} = \begin{bmatrix} 0 \\ -5 \end{bmatrix}$ onto $\boldsymbol{w}$. $\boldsymbol{p} =$ ____________

b) Use command **project(u,w)** to determine the quadrant in which $\boldsymbol{p}$ is located.

Record the quadrant here. ____________

c) Find a scalar k for which the vector $k\boldsymbol{p}$ has a norm that is equal to one.

$k =$ ____________

d) Find a vector $\boldsymbol{s}$ that is orthogonal to $\mathbf{span}\{\boldsymbol{w}\}$. $\boldsymbol{s} =$ ____________

3. Let $\boldsymbol{w} = \begin{bmatrix} 3 \\ 4 \end{bmatrix}$.

a) Find the projection $\boldsymbol{p}$ of $\boldsymbol{u} = \begin{bmatrix} 4 \\ -3 \end{bmatrix}$ onto $\boldsymbol{w}$. $\boldsymbol{p} =$ ____________

b) Briefly explain your answer to part a). Hint: Use command **project(u,w)**.

____

4. Let $\boldsymbol{w1} = \begin{bmatrix} 1 \\ 2 \\ 0 \end{bmatrix}$, $\boldsymbol{w2} = \begin{bmatrix} 2 \\ -1 \\ 3 \end{bmatrix}$, and $\boldsymbol{u} = \begin{bmatrix} 1 \\ 4 \\ 1 \end{bmatrix}$.

a) Find the projection $\boldsymbol{p}$ of $\boldsymbol{u}$ onto $\mathbf{span}\{\boldsymbol{w1}, \boldsymbol{w2}\}$. $\boldsymbol{p} =$ ____________

b) Find a vector $\boldsymbol{s}$ that is orthogonal to $\mathbf{span}\{\boldsymbol{w1}, \boldsymbol{w2}\}$. $\boldsymbol{s} =$ ____________

5. Let $\boldsymbol{w1} = \begin{bmatrix} 5 \\ 2 \\ -3 \end{bmatrix}$, $\boldsymbol{w2} = \begin{bmatrix} 1 \\ -1 \\ 1 \end{bmatrix}$, and $\boldsymbol{u} = \begin{bmatrix} 1 \\ 1 \\ 0 \end{bmatrix}$.

a) Find the projection $\boldsymbol{p}$ of $\boldsymbol{u}$ onto $\boldsymbol{W} = \mathbf{span}\{\boldsymbol{w1}, \boldsymbol{w2}\}$. $\boldsymbol{p} =$ ____________

b) What is the relationship between $\mathbf{proj}_{\boldsymbol{W}}\boldsymbol{u}$ and $\mathbf{proj}_{\boldsymbol{w}_1}\boldsymbol{u}$?
Explain why it occurs.

____

6. Let $\boldsymbol{w1} = \begin{bmatrix} \sqrt{2}/2 \\ 0 \\ \sqrt{2}/2 \end{bmatrix}$, $\boldsymbol{w2} = \begin{bmatrix} -\sqrt{2}/2 \\ 0 \\ \sqrt{2}/2 \end{bmatrix}$ and $\boldsymbol{W} = \mathbf{span}\{\boldsymbol{w1}, \boldsymbol{w2}\}$.

a) Show that $\{\boldsymbol{w1}, \boldsymbol{w2}\}$ is an orthonormal set.

b) Use 10.4 to determine $\mathbf{proj}_{\boldsymbol{W}}\boldsymbol{u}$ where $\boldsymbol{u} = \begin{bmatrix} 4 \\ 2 \\ 1 \end{bmatrix}$.

7. Let $\boldsymbol{w1} = \begin{bmatrix} \sqrt{3}/3 \\ \sqrt{3}/3 \\ \sqrt{3}/3 \end{bmatrix}$, $\boldsymbol{w2} = \begin{bmatrix} -\sqrt{2}/2 \\ 0 \\ \sqrt{2}/2 \end{bmatrix}$ and $\boldsymbol{W} = \mathbf{span}\{\boldsymbol{w1}, \boldsymbol{w2}\}$.

a) Show that $\{\boldsymbol{w1}, \boldsymbol{w2}\}$ is an orthonormal set.

b) Use 10.4 to determine $\mathbf{proj}_{\boldsymbol{W}}\boldsymbol{u}$ where $\boldsymbol{u} = \begin{bmatrix} 4 \\ 2 \\ 1 \end{bmatrix}$.

Section 10.3

The Gram – Schmidt Process

The ideas about projections in Section 10.2 actually tell us a way to construct an orthonormal basis from an existing basis provided we build the new basis one vector at a time.

The Gram-Schmidt process takes a basis $\mathbf{S} = \{\boldsymbol{u_1}, \boldsymbol{u_2}, ..., \boldsymbol{u_n}\}$ for a subspace of an inner product space $\mathbf{V}$ and produces a new basis $\mathbf{T} = \{\boldsymbol{w_1}, \boldsymbol{w_2}, ..., \boldsymbol{w_n}\}$ whose vectors form an orthonormal set. The process is often performed in two stages:

- First from the $\mathbf{S}$-basis generate a basis $\{\boldsymbol{v_1}, \boldsymbol{v_2}, ..., \boldsymbol{v_n}\}$ of vectors that are mutually orthogonal. That is, $(\boldsymbol{v_i}, \boldsymbol{v_j}) = 0$, $i \neq j$.

- Second normalize each of the orthogonal basis vectors into a unit vector.

The first stage involves solving a set of equations and the second is easily performed using $\boldsymbol{w_i} = \boldsymbol{v_i} / \parallel \boldsymbol{v_i} \parallel$. At each step in the first stage we use projections onto subspaces.

The First Stage

Step 1. Define $\boldsymbol{v_1} = \boldsymbol{u_1}$.

Step 2. Look for a vector $\boldsymbol{v_2}$ in the $\mathbf{span}\{\boldsymbol{v_1}, \boldsymbol{u_2}\}$ that is orthogonal to $\boldsymbol{v_1}$. This will then guarantee that

$$\begin{aligned} \mathbf{span}\{\boldsymbol{u_1}, \boldsymbol{u_2}\} &= \mathbf{span}\{\boldsymbol{v_1}, \boldsymbol{u_2}\} && \text{since } \boldsymbol{v_1} = \boldsymbol{u_1} \\ &= \mathbf{span}\{\boldsymbol{v_1}, \boldsymbol{v_2}\} && \text{since } \boldsymbol{v_2} \text{ is a linear combination of } \boldsymbol{v_1} \text{ and } \boldsymbol{u_2} \end{aligned}$$

Let $\boldsymbol{v_2} = k_1\boldsymbol{v_1} + k_2\boldsymbol{u_2}$. Find k_1 and k_2 so that $(\boldsymbol{v1}, \boldsymbol{v2}) = \mathbf{0}$.

$$0 = (\boldsymbol{v_1}, \boldsymbol{v_2}) = k_1(\boldsymbol{v_1}, \boldsymbol{v_1}) + k_2(\boldsymbol{v_1}, \boldsymbol{u_2})$$

We have one equation in two unknowns, so let $k_2 = 1$ and solve for k_1. We get

$$k_1 = \frac{-(\boldsymbol{v_1}, \boldsymbol{u_2})}{(\boldsymbol{v_1}, \boldsymbol{v_1})}$$

thus we have

$$\boldsymbol{v_2} = \boldsymbol{u_2} - \frac{(\boldsymbol{v_1}, \boldsymbol{u_2})}{(\boldsymbol{v_1}, \boldsymbol{v_1})}\boldsymbol{v_1} = \boldsymbol{u_2} - \mathbf{proj}_{\boldsymbol{v_1}}\, \boldsymbol{u_2}$$

Step 3. Look for a vector $\boldsymbol{v_3}$ in $\mathbf{span}\{\boldsymbol{v_1}, \boldsymbol{v_2}, \boldsymbol{u_3}\}$ that is orthogonal to both $\boldsymbol{v_1}$ and $\boldsymbol{v_2}$. This will guarantee that $\mathbf{span}\{\boldsymbol{u_1}, \boldsymbol{u_2}, \boldsymbol{u_3}\} = \mathbf{span}\{\boldsymbol{v_1}, \boldsymbol{v_2}, \boldsymbol{u_3}\} = \mathbf{span}\{\boldsymbol{v_1}, \boldsymbol{v_2}, \boldsymbol{v_3}\}$. Let $\boldsymbol{v_3} = k_1\boldsymbol{v_1} + k_2\boldsymbol{v_2} + k_3\boldsymbol{u_3}$. Find k_1, k_2, and k_3 so that $(\boldsymbol{v_1}, \boldsymbol{v_3}) = \mathbf{0}$ and $(\boldsymbol{v_2}, \boldsymbol{v_3}) = 0$.

$$\begin{aligned} 0 &= (\boldsymbol{v_1}, \boldsymbol{v_3}) = \boldsymbol{k_1}(\boldsymbol{v_1}, \boldsymbol{v_1}) + \boldsymbol{k_2}(\boldsymbol{v_1}, \boldsymbol{v_2}) + \boldsymbol{k_3}(\boldsymbol{v_1}, \boldsymbol{u_3}) \\ 0 &= (\boldsymbol{v_2}, \boldsymbol{v_3}) = \boldsymbol{k_1}(\boldsymbol{v_2}, \boldsymbol{v_1}) + \boldsymbol{k_2}(\boldsymbol{v_2}, \boldsymbol{v_2}) + \boldsymbol{k_3}(\boldsymbol{v_2}, \boldsymbol{u_3}) \end{aligned}$$

Since by construction $(\boldsymbol{v1}, \boldsymbol{v2}) = 0$ the preceding equations simplify to

$$\begin{aligned} k_1(\boldsymbol{v_1}, \boldsymbol{v_1}) \qquad\qquad\qquad & + k_3(\boldsymbol{v_1}, \boldsymbol{u_3}) = 0 \\ k_2(\boldsymbol{v_2}, \boldsymbol{v_2}) & + k_3(\boldsymbol{v_2}, \boldsymbol{u_3}) = 0 \end{aligned}$$

Thus we have 2 equations in 3 unknowns. Let $k_3 = 1$, then we find that

$$k_1 = \frac{-(\boldsymbol{v_1}, \boldsymbol{u_3})}{(\boldsymbol{v_1}, \boldsymbol{v_1})} \qquad \text{and} \qquad k_2 = \frac{-(\boldsymbol{v_2}, \boldsymbol{u_3})}{(\boldsymbol{v_2}, \boldsymbol{v_2})}$$

and hence

$$\boldsymbol{v_3} = \boldsymbol{u_3} - \frac{(\boldsymbol{v_1}, \boldsymbol{u_3})}{(\boldsymbol{v_1}, \boldsymbol{v_1})}\boldsymbol{v_1} - \frac{(\boldsymbol{v_2}, \boldsymbol{u_3})}{(\boldsymbol{v_2}, \boldsymbol{v_2})}\boldsymbol{v_2} = \boldsymbol{u_3} - \mathbf{proj}_{\mathbf{span}\{\boldsymbol{v_1}, \boldsymbol{v_2}\}}\, \boldsymbol{u_3}$$

Other steps: $\boldsymbol{v_k} = \boldsymbol{u_k} - \mathbf{proj}_{\mathbf{span}\{\boldsymbol{v_1}, \boldsymbol{v_2}, \ldots, \boldsymbol{v_{k-1}}\}}\, \boldsymbol{u_k}$

The Second Stage

The orthonormal basis for $\mathbf{V}$ is given by

$$\{\boldsymbol{w_1}, \boldsymbol{w_2}, \ldots, \boldsymbol{w_n}\} = \left\{ \frac{\boldsymbol{v_1}}{\|\boldsymbol{v_1}\|}, \frac{\boldsymbol{v_2}}{\|\boldsymbol{v_2}\|}, \ldots, \frac{\boldsymbol{v_n}}{\|\boldsymbol{v_n}\|} \right\}$$

Example 1. Let $\mathbf{V}=\mathbf{span}\{\boldsymbol{u}_1, \boldsymbol{u}_2, \boldsymbol{u}_3\}$ where

$$\boldsymbol{u}_1 = \begin{bmatrix} 2 \\ 1 \\ 0 \\ 4 \end{bmatrix}, \boldsymbol{u}_2 = \begin{bmatrix} 1 \\ 0 \\ 1 \\ 3 \end{bmatrix}, \boldsymbol{u}_3 = \begin{bmatrix} 1 \\ 2 \\ 1 \\ 0 \end{bmatrix}$$

Use the Gram-Schmidt process to find an orthonormal basis for $\mathbf{V}$.

Step 1. Define $\boldsymbol{v}_1 = \boldsymbol{u}_1 = \begin{bmatrix} 2 \\ 1 \\ 0 \\ 4 \end{bmatrix}$.

Step 2. Compute $\boldsymbol{v}_2 = \boldsymbol{u}_2 - \mathbf{proj}_{\boldsymbol{v}_1} \boldsymbol{u}_2 = \boldsymbol{u}_2 - \frac{(\boldsymbol{v}_1,\boldsymbol{u}_2)}{(\boldsymbol{v}_1,\boldsymbol{v}_1)}\boldsymbol{v}_1 = \begin{bmatrix} 1 \\ 0 \\ 1 \\ 3 \end{bmatrix} - \frac{14}{21}\begin{bmatrix} 2 \\ 1 \\ 0 \\ 4 \end{bmatrix} = \begin{bmatrix} -\frac{1}{3} \\ -\frac{2}{3} \\ 1 \\ \frac{1}{3} \end{bmatrix}$.

Step 3. Compute $\boldsymbol{v}_3 = \boldsymbol{u}_3 - \mathbf{proj}_{\mathbf{span}\{\boldsymbol{v}_1, \boldsymbol{v}_2\}} \boldsymbol{u}_3 = \boldsymbol{u}_3 - \frac{(\boldsymbol{v}_1,\boldsymbol{u}_3)}{(\boldsymbol{v}_1,\boldsymbol{v}_1)}\boldsymbol{v}_1 - \frac{(\boldsymbol{v}_2,\boldsymbol{u}_3)}{(\boldsymbol{v}_2,\boldsymbol{v}_2)}\boldsymbol{v}_2 =$

$$\begin{bmatrix} 1 \\ 2 \\ 1 \\ 0 \end{bmatrix} - \frac{4}{21}\begin{bmatrix} 2 \\ 1 \\ 0 \\ 4 \end{bmatrix} - \frac{(-2/3)}{15/9}\begin{bmatrix} \frac{-1}{3} \\ \frac{-2}{3} \\ 1 \\ \frac{1}{3} \end{bmatrix} = \begin{bmatrix} \frac{17}{35} \\ \frac{54}{35} \\ \frac{7}{5} \\ \frac{-22}{35} \end{bmatrix}.$$

The set $\{\boldsymbol{v}_1, \boldsymbol{v}_2, \boldsymbol{v}_3\}$ is an orthogonal basis for $\mathbf{V}$. An orthonormal basis is obtained by dividing each vector by it length.

$$\boldsymbol{w}_1 = \frac{\boldsymbol{v}_1}{\sqrt{21}}, \qquad \boldsymbol{w}_2 = \frac{\boldsymbol{v}_2}{\sqrt{17/9}}, \qquad \boldsymbol{w}_3 = \frac{\boldsymbol{v}_3}{\sqrt{6090/1225}}$$

For $\mathbf{V} = R^n$ and the standard inner product both stages of the Gram-Schmidt process are available in MATLAB routine **gschmidt**. Type **help gschmidt** for more details. The following examples illustrate the use of routine **gschmidt**.

Example 2. Let $\mathbf{S} = \{\boldsymbol{u}_1, \boldsymbol{u}_2, \boldsymbol{u}_3\} = \left\{ \begin{bmatrix} 1 \\ 0 \\ 2 \end{bmatrix}, \begin{bmatrix} 2 \\ 1 \\ 0 \end{bmatrix}, \begin{bmatrix} 0 \\ 2 \\ 1 \end{bmatrix} \right\}$ be a basis for R^3. To find an orthonormal basis from $\mathbf{S}$ using MATLAB enter the vectors $\boldsymbol{u}_1, \boldsymbol{u}_2, \boldsymbol{u}_3$ as columns of a matrix $\boldsymbol{A}$ and type

B = gschmidt(A)

The display generated is

```
B =

    0.4472    0.7807   -0.4364
         0    0.4880    0.8729
    0.8944   -0.3904    0.2182
```

The columns of $\boldsymbol{B}$ are an orthonormal basis for R^3.

Example 3. We will show how to find an orthonormal basis for R^4 containing scalar multiples of the vectors

$$\boldsymbol{v_1} = \begin{bmatrix} 1 \\ 0 \\ 1 \\ -1 \end{bmatrix} \text{ and } \boldsymbol{v_2} = \begin{bmatrix} -1 \\ 1 \\ 2 \\ 1 \end{bmatrix}.$$

First enter $\boldsymbol{v_1}$ and $\boldsymbol{v_2}$ into MATLAB as vectors **v1** and **v2**, respectively. To find a basis containing scalar multiples of $\boldsymbol{v_1}$ and $\boldsymbol{v_2}$, use commands

A = [v1 v2 eye(4)]
rref(A)

The display indicates that the first four columns of $\boldsymbol{A}$ form a basis for R^4. The command **S = A(:,1:4)** produces the matrix with those columns. Type the command

T = gschmidt(S)

The display is

```
T =
    0.5774   -0.3780    0.7237         0
         0    0.3780    0.1974    0.9045
    0.5774    0.7559   -0.0658   -0.3015
   -0.5774    0.3780    0.6580   -0.3015
```

Column 1 of $\boldsymbol{T}$ is $\left(\frac{1}{\|\boldsymbol{v_1}\|}\right)\boldsymbol{v_1}$ and column 2 of $\boldsymbol{T}$ is $\left(\frac{1}{\|\boldsymbol{v_2}\|}\right)\boldsymbol{v_2}$, hence the columns of $\boldsymbol{T}$ form the desired orthonormal basis for R^4.

Explain what to do if **rref(A)** did not indicate that the first four columns of $\boldsymbol{A}$ form a basis for R^4.

Exercises 10.3

1. Let $\mathbf{V} = R^3$ with the standard inner product and let

$$\mathbf{S} = \{\boldsymbol{u}_1, \boldsymbol{u}_2, \boldsymbol{u}_3\} = \left\{ \begin{bmatrix} 1 \\ 2 \\ 0 \end{bmatrix}, \begin{bmatrix} 1 \\ 0 \\ 0 \end{bmatrix}, \begin{bmatrix} 1 \\ 0 \\ 1 \end{bmatrix} \right\}.$$

Use routine **gschmidt** in MATLAB to obtain an orthonormal basis **T** and then find the coordinates of $\boldsymbol{x} = \begin{bmatrix} 1 \\ 2 \\ 3 \end{bmatrix}$ relative to **T**. Record the orthonormal basis and the coordinates of $\boldsymbol{x}$ below.

2. Let $\mathbf{V} = R^4$ with the standard inner product and let

$$\mathbf{S} = \{\boldsymbol{u}_1, \boldsymbol{u}_2, \boldsymbol{u}_3, \boldsymbol{u}_4\} = \left\{ \begin{bmatrix} -1 \\ 2 \\ 0 \\ 1 \end{bmatrix}, \begin{bmatrix} 2 \\ 1 \\ 1 \\ 0 \end{bmatrix}, \begin{bmatrix} 0 \\ 1 \\ 0 \\ 1 \end{bmatrix}, \begin{bmatrix} 1 \\ 1 \\ 0 \\ 1 \end{bmatrix} \right\}.$$

Use routine **gschmidt** in MATLAB to obtain an orthonormal basis **T** and then find the coordinates of $\boldsymbol{x} = \begin{bmatrix} 4 \\ 0 \\ 2 \\ 1 \end{bmatrix}$ relative to **T**. Record the orthonormal basis and the coordinates of $\boldsymbol{x}$ below.

3. Let $\mathbf{V} = R^4$ with the standard inner product and let

$$\mathbf{S} = \{\boldsymbol{u_1}, \boldsymbol{u_2}, \boldsymbol{u_3}, \boldsymbol{u_4}\} = \left\{ \begin{bmatrix} .5 \\ .5 \\ .5 \\ .5 \end{bmatrix}, \begin{bmatrix} .5 \\ .5 \\ -.5 \\ -.5 \end{bmatrix}, \begin{bmatrix} .5 \\ -.5 \\ -.5 \\ .5 \end{bmatrix}, \begin{bmatrix} .5 \\ -.5 \\ .5 \\ -.5 \end{bmatrix} \right\}.$$

a) Is **S** an orthonormal basis? Circle one: Yes No

Explain your answer.

b) In MATLAB form the matrix $\boldsymbol{T}$ whose columns are the vectors in **S**. Generate a random vector in R^4 using command **x = rand(4,1)** and then compute $\| \boldsymbol{x} \|$ and $\| \boldsymbol{Tx} \|$. How are the values of the norms related? Repeat the experiment for another arbitrary vector.

____

____

4. Let $\boldsymbol{v_1} = \begin{bmatrix} 1 \\ 2 \\ 2 \end{bmatrix}$ and $\boldsymbol{v_2} = \begin{bmatrix} 1 \\ 3 \\ 1 \end{bmatrix}$. In MATLAB form the matrix **A = [v1 v2]** and then use command **gschmidt(A)**. Explain the meaning of the display generated.

____

____

5. Let $\boldsymbol{A} = \begin{bmatrix} 1 & i & 0 \\ i & 0 & 1 \end{bmatrix}$.

a) In MATLAB use command **A′**. Record the result. **A′** = ______________

b) In MATLAB use command **C = A′∗A**. Record the result. **C** = ______________

c) What is the relation between $\boldsymbol{C}$ and $\boldsymbol{C'}$?

____

d) Experiment with other complex matrices $\boldsymbol{A}$ to confirm or reject your answer in part c).

Circle one: confirmed not confirmed.

6. A complex matrix $\boldsymbol{A}$ is called Hermitian if it is equal to its conjugate transpose. The command $\mathbf{A}'$ gives the conjugate transpose in MATLAB .

a) How can you use MATLAB to determine if the matrix $\boldsymbol{A}$ below is Hermitian?

$$\boldsymbol{A} = \begin{bmatrix} 2 & 3-3i \\ 3+3i & 5 \end{bmatrix}$$

____

b) Compute $\boldsymbol{r} = \boldsymbol{x}' * \boldsymbol{A} * \boldsymbol{x}$ for the complex vector below.

$$\boldsymbol{x} = \begin{bmatrix} i \\ 1-i \end{bmatrix} \qquad \boldsymbol{r} = \underline{\hspace{4cm}}$$

Is $\boldsymbol{r}$ a real number? (Circle one:) YES NO

c) Experiment with other complex vectors $\boldsymbol{x}$ to determine whether $\boldsymbol{x}'\boldsymbol{A}\boldsymbol{x}$ will always be a real number. (Circle one:)

Always a real number for this matrix $\boldsymbol{A}$. Not always a real number.

d) Experiment with another Hermitian matrix $\boldsymbol{A}$ and arbitrary vector $\boldsymbol{x}$ to see if $\boldsymbol{r} = \boldsymbol{x}' * \boldsymbol{A} * \boldsymbol{x}$ is always a real number.

(Circle one:) Always a real number. Not always a real number.

7. Let $\mathbf{V} = R^4$ with the standard inner product and let

$$\boldsymbol{v_1} = \begin{bmatrix} 3 \\ 1 \\ 2 \\ 0 \end{bmatrix}, \qquad \boldsymbol{v_2} = \begin{bmatrix} 1 \\ -1 \\ -1 \\ 1 \end{bmatrix}, \qquad \boldsymbol{v_3} = \begin{bmatrix} 0 \\ -2 \\ 1 \\ -1 \end{bmatrix}.$$

a) Find an orthonormal basis for R^4 containing scalar multiples of the vectors $\boldsymbol{v_1}$ and $\boldsymbol{v_2}$. Record your result below.

b) Find an orthonormal basis for R^4 containing scalar multiples of the vectors $\boldsymbol{v_1}$, $\boldsymbol{v_2}$, $\boldsymbol{v_3}$. Record your result below.

<< NOTES; COMMENTS; IDEAS >>

LAB 11

Plane Linear Transformations

Topics: plane linear transformations (reflections, rotations, compressions, expansions, shears), geometric displays; routines **planelt and matrixmaps**.

Introduction

Linear transformations play a central role in linear algebra. This lab presents routines that provide a geometric visualization of the underlying concepts for linear transformations in the plane. No prior knowledge of linear transformations is assumed. Indeed, the geometric approach motivates many of the concepts that are fundamental to all linear transformations.

In Section 11.1 the routine **planelt** makes explicit use of the matrix representation of a linear transformation. Here the user selects a linear transformation from a menu of plane linear transformations and applies it to well-known figures from high school geometry. The linear transformations include rotations, reflections, expansions/compressions, and shears. The figures include a square, rectangle, parallelogram, triangle, and pentagon. Users can define their own figures and plane linear transformations.

A series of experiments anticipates properties of a linear transformation T. It is emphasized that sometimes the geometric nature of T addresses the property more directly, whereas at other times the algebraic nature, that is, computations involving the matrix representation, provides a better route. This approach transcends linear algebra. It goes to the heart of mathematics: specific examples versus theorems, proofs versus counterexamples.

Planelt provides a visualization of other basic concepts as well. By defining a linear transformation in terms of a singular (non-invertible) matrix, it is possible to examine the dimension of a space, the rank of a matrix, and the kernel of a linear transformation.

In Section 11.2 we use another routine, **matrixmaps**, to explore further connections among the topics of linear transformations, matrix representations, determinants, and composite functions based on linear transformations. In this section we introduce the notion of homogeneous coordinates so that translations of plane figures can be performed by matrix multiplication. Experiments are included to provide a physical meaning to the value of the determinant of a matrix representing a linear transformation.

Section 11.1

Graphics Experiments

This section develops the idea that a matrix representation (a concept in algebra) accompanies each linear transformation (a concept in geometry). The interplay between an algebraic approach and a geometric approach is particularly beneficial when learning mathematics. The geometry provides a visualization of the algebra.

We investigate properties of certain plane linear transformations. The main tool is routine **planelt**, which views a geometric figure in three separate frames: present, prior and original. In MATLAB type command **planelt**. Read the Introduction and the General Directions. Next comes a menu of Figure Choices containing a list of geometric figures. Selections are made by typing the appropriate number and then pressing ENTER. To begin, choose a triangle (option 4). This produces another menu. For now choose option number 2, *'Use this figure. Go to select transformations.'* This in turn brings up a menu of Plane Linear Transformations. There are four special types:

rotations
reflections
expansions/compressions
shears

All of these transformations should be familiar except, perhaps, shears.

Four options on the menu of Plane Linear Transformations that we will use frequently are different from those in which you select a plane linear transformation. Option 10 allows you to enter your own 2×2 matrix. This option defines a plane linear transformation in terms of that matrix. We will make use of it frequently. The other three choices are option 11, which restores the original figure, option 0, which starts with a new figure, and option -1, which is used to correct mistakes made in the previous selection.

Matrix Representations

We begin by reflecting the triangle about the x-axis (select option 2.) Let T denote this transformation. The Current Figure shows the result of the reflection on the Previous Figure. Here the Previous Figure is the same as the Original Figure. The matrix in the lower left quadrant of the screen displays the matrix representation $\boldsymbol{M_T}$ of linear transformation $\boldsymbol{T}$. Here

$$\boldsymbol{M_T} = \begin{bmatrix} 1 & 0 \\ 0 & -1 \end{bmatrix}.$$

The linear transformation T 'moves around' the points that are vertices of the triangle. This triangle has vertices as depicted in Figure 1.

The image of the vertex $\boldsymbol{v} = \begin{bmatrix} 0 \\ 2 \end{bmatrix}$ is

$$\boldsymbol{M_T} \cdot \boldsymbol{v} = \begin{bmatrix} 1 & 0 \\ 0 & -1 \end{bmatrix} \begin{bmatrix} 0 \\ 2 \end{bmatrix} = \begin{bmatrix} 0 \\ -2 \end{bmatrix}.$$

$\boldsymbol{T}$ also moves points on the triangle's edges. For instance the image of $\boldsymbol{w} = \begin{bmatrix} 1 \\ 1 \end{bmatrix}$ is

$$\boldsymbol{M_T} \cdot \boldsymbol{w} = \begin{bmatrix} 1 \\ -1 \end{bmatrix}$$

Notice that $\text{T}(\boldsymbol{v}) = \boldsymbol{M_T v}$ for every such point $\boldsymbol{v}$. It turns out that $\text{T}(\boldsymbol{v}) = \boldsymbol{M_T v}$ for every column vector $\boldsymbol{v}$. This formula captures what it means for $\boldsymbol{M_T}$ to be the matrix representation of T. You may find it useful to regard $\boldsymbol{M_T}$ as a kind of code for T, carrying information in much

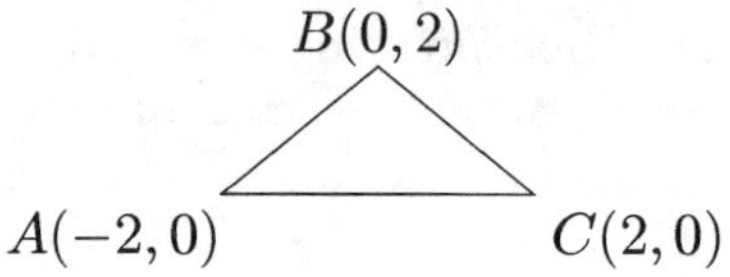

Figure 1

the same way that DNA carries the genetic code.

Composition of Transformations

Next we use the composition of functions. Recall that if $f(x)$ and $g(x)$ are functions then $g \circ f$ is the function defined by $(g \circ f)(x) = g(f(x))$. We will denote transformations by upper case letters like S and T, and vectors by lower case letters like $\boldsymbol{v}$, so the definition becomes

$$(\mathrm{S}{\circ}\mathrm{T})(\boldsymbol{v}) = \mathrm{S}(\mathrm{T}(\boldsymbol{v})).$$

The order of the terms in the composition is vital here: T comes first and S second.

We have defined T to be reflection about the x-axis. The screen should contain the result of T being applied to a triangle. Now we apply another plane linear transformation to the Current Figure. So press ENTER. Then rotate the figure 60° counterclockwise (option 1 with angle 60). Let S be the linear transformation 'rotation by 60°.' The matrix representation for S is

$$\boldsymbol{M}_S = \begin{bmatrix} 0.5 & -.866 \\ .866 & 0.5 \end{bmatrix}$$

Sketch the Current Figure, which we will label (I):

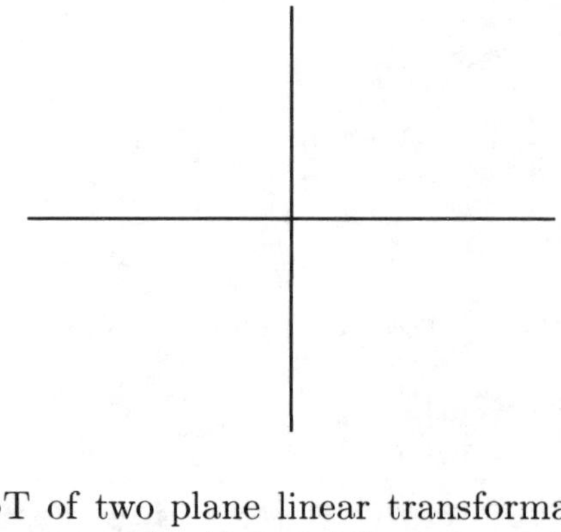

(I)

The composition S∘T of two plane linear transformations is itself a plane linear transformation. We seek a way to write the matrix representation of S∘T in terms of the matrix representations of S and T. Example 1 examines a particular case.

Example 1. Let T be reflection about the x-axis and S be rotation by $60°$. We will determine what matrix corresponds to $S \circ T$, where the notation means that a figure is reflected first and rotated second. Let $\boldsymbol{A}$ be the product of the matrix representations $\boldsymbol{M}_T$ and $\boldsymbol{M}_S$,

$$\boldsymbol{A} = \boldsymbol{M}_S \cdot \boldsymbol{M}_T = \begin{bmatrix} 0.5 & -0.866 \\ 0.866 & 0.5 \end{bmatrix} \begin{bmatrix} 1 & 0 \\ 0 & -1 \end{bmatrix} = \begin{bmatrix} 0.5 & 0.866 \\ 0.866 & -0.5 \end{bmatrix}.$$

Now restore the triangle and for the transformation select option 10. ('Use your 2×2 matrix'.) Follow the directions there for entering the matrix $\boldsymbol{A}$. Record the Current Figure, which we label II:

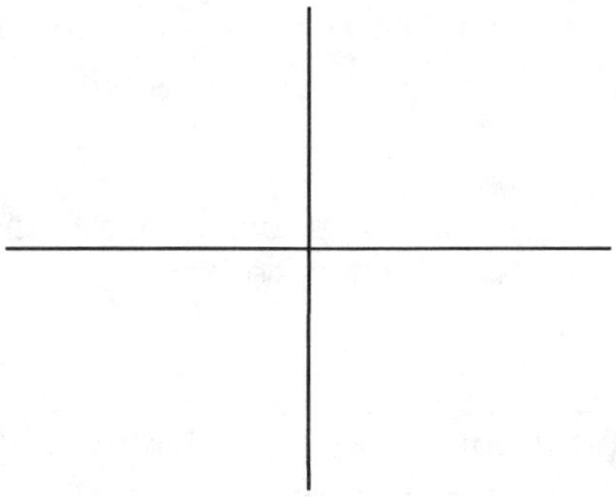

(II)

Notice that Figures I and II are identical in size and orientation. This means that the two transformations, one defined by the composition S∘T and the other by the matrix $\boldsymbol{A}$, have the same effect on the triangle.

Now test if the two transformations have the same effect on another figure. To enter a new figure select option 0. Then select the pentagon (option 5). Once again reflect this figure about the x-axis and then rotate it by $60°$. Record the Current Figure in III.

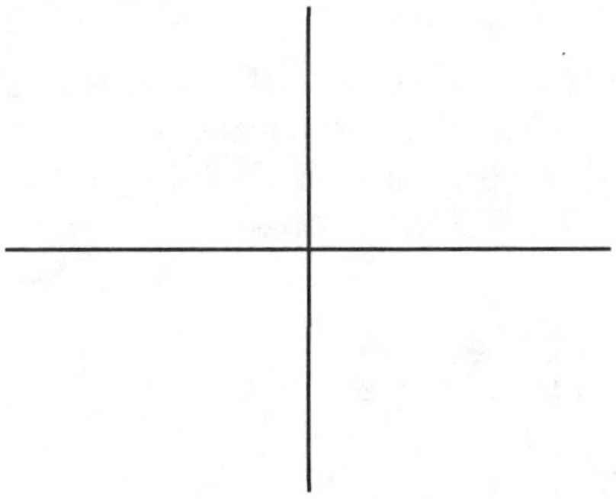

(III)

Next restore the original figure (option 11) and perform the transformation whose matrix representation is $\boldsymbol{A}$. Record the Current Figure in IV. The two current figures are identical in size and orientation, thus attesting to the fact that, for the pentagon, the transformation whose matrix representation is $\boldsymbol{A}$ and is equal to the composition of the two transformations, $S \circ T$. It turns out that this equality holds for every figure in the plane, and hence it holds for every point in the plane.

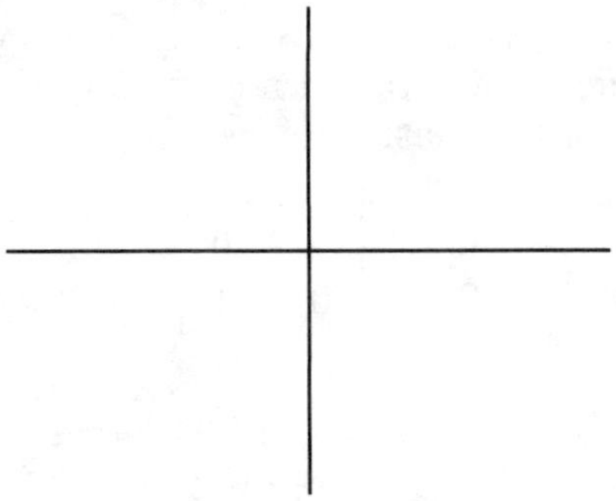

(IV)

We thus draw the conclusion that **the matrix representation of $S \circ T$ is equal to the matrix representation of S times the matrix representation of T in that order.**

The result in Example 1 holds not only for the plane linear transformations defined there but for any pair of linear transformations. This fact is so important that we will highlight it.

> If S and T are linear transformations with matrix representations $\boldsymbol{M}_S$ and $\boldsymbol{M}_T$, respectively, then the matrix representation of the linear transformation $S \circ T$ is $\boldsymbol{M}_S \cdot \boldsymbol{M}_T$.

Properties of Composition

We now examine some properties of linear transformations using either an algebraic approach or a geometric approach, whichever suits our purpose at the time. We compare each property of transformations under composition with the same property of matrices under multiplication. Let's begin with the commutative law.

In general multiplication of matrices is **not** commutative. Is the composition of linear transformations commutative? One way to answer this question is to proceed by experimentation. Once again, let T be reflection about the x-axis and S be rotation by 60°. Restore the original figure (the pentagon) and perform $T \circ S$, which rotates first and reflects second. Record the current Figure in V.

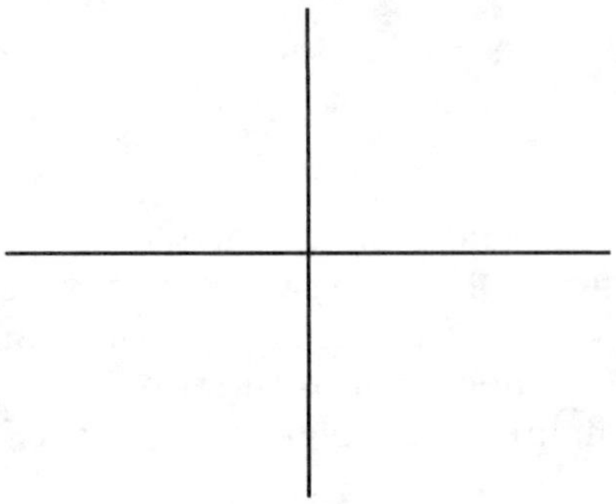

(V)

Since figures IV and V are not identical it follows that $S \circ T \neq T \circ S$. Therefore, in general, transformations do not commute under composition. This is analogous to the behavior of matrix multiplication.

Next we exploit the relationship between matrices under multiplication and transformations under composition to investigate inverses. The transformation that corresponds to the identity matrix is the identity mapping I, which maps each vector to itself. Recall that a square matrix is invertible if it has an inverse. Similarly, a transformation T is invertible if there is a transformation S with the property that $T \circ S = I$ and $S \circ T = I$. The transformation S is called the inverse of T and is usually denoted by T^{-1}.

To visualize this situation, start with a new figure, the parallelogram (option 3). Let T be rotation by 45° and S be rotation by 315°. Perform T then perform S. Notice that the current figure is identical to the Original Figure. Next restore the parallelogram and perform S first and then T. Once again the Current Figure is identical to the Original Figure. This suggests that $T \circ S = S \circ T = I$ so $S = T^{-1}$. What is $\boldsymbol{M}_T \boldsymbol{M}_S$? How are $\boldsymbol{M}_T$ and $\boldsymbol{M}_S$ related?

______________________________ ______________________________

Now start with a new figure, the 2x1 rectangle (option 2.) Expand it in the x-direction by $k = 2$. Next choose a transformation so that the resulting Current Figure is identical to the Original Figure. Record the transformation you chose. Now start with the triangle and expand it in the x-direction by $k = 2$. Apply the transformation you chose previously. The result is identical to the Original Figure. Using the previous results, state a conjecture for the inverse of expansion in the x-direction by $k > 1$.

____

What do you think is the inverse of compression in the x-direction by $k < 1$?

____

Exercises 11.1

1. Experiment with several figures to see if $S \circ T = T \circ S$ for the following choices of S and T. Record your conclusions.

a) T = reflect about y-axis
S = expand/compress in x-direction with $k = 2/3$

____

b) T = shear in y-direction with $k = 2$
S = reflect about line $y = x$

____

2. Respond to each of the following in the space provided.

a) What is the matrix representation $\boldsymbol{M}_T$ of reflection T about the line $y = -x$?

b) What is the matrix representation $\boldsymbol{M}_S$ of shear S in the x-direction with $k = 2$?

c) Write a formula for the matrix representation of $T \circ S$ in terms of $\boldsymbol{M}_T$ and $\boldsymbol{M}_S$.

d) What is the matrix representation of $T \circ S$?

3. Let $\boldsymbol{M}_T$ be the matrix representation of reflection T about $y = x$, $\boldsymbol{M}_S$ be the matrix representation of expansion S in the y-direction with $k = 3$, and $\boldsymbol{M}_R$ be the matrix representation of rotation R by 225°.

a) Write a formula for the matrix representation of $T \circ S \circ R$ in terms of $\boldsymbol{M}_T$, $\boldsymbol{M}_S$, and $\boldsymbol{M}_R$.

b) What is the matrix representation of $T \circ S \circ R$?

4. Let T and S be the transformations whose matrix representations are $\boldsymbol{M}_T = \begin{bmatrix} 1 & 2 \\ 3 & 4 \end{bmatrix}$

and $\boldsymbol{M}_S = \begin{bmatrix} 2 & 0 \\ 0 & 2 \end{bmatrix}$. Is it true that $S \circ T = T \circ S$? Why? (Respond below.)

5. Let T be reflection about $y = x$ and S be reflection about $y = -x$. Experiment with several figures to determine the nature of the transformation $T \circ S$. Record your result by describing the image of a point (x, y).

6. Respond to each of the following in the space provided.

a) What is the matrix representation $\boldsymbol{M}_T$ of compression T in the y-direction with factor 0.2?

b) What is the matrix representation of T^{-1}?

7. Respond to each of the following in the space provided.

a) What is the matrix representation $\boldsymbol{M}_T$ of shear T in the y-direction with $k = 2/3$?

b) What is the matrix representation of T^{-1}?

c) Describe in words the transformation T^{-1}.

8. How is the matrix representation of T^{-1} related to the matrix representation of T? (See Exercises 6 and 7.)

Section 11.2

More Graphics Experiments

Introduction

The mathematics underlying computer graphics is closely connected to matrix multiplication as we saw in Section 11.1. We perform rotations, reflections, or scaling operations using a square matrix $\boldsymbol{A}$ of size 2×2. The operations are characterized as a function f which works on a *'picture'*. viewed as a set of data, to produce an *'image'*.

$$f('picture') = 'image'$$

When the data representing the *'picture'* is properly arranged the operation of function f is executed by a matrix $\boldsymbol{A}$:

$$f('picture') = \boldsymbol{A} * 'picture' = 'image'.$$

Unfortunately a general transformation not only includes rotations, reflections, and scalings, but also translations or shifts (which "carry" the picture without distortion beyond where it used to be to a new place), which cannot be expressed using a transformation matrix of corresponding size 2×2. To use matrix multiplication seamlessly for translations we introduce homogeneous coordinates[1], which are an extended representation of the data representing the *'picture'*.

Recall that common plane transformations are associated with 2×2 matrices as shown below.

▷ Identity transformation: $\begin{bmatrix} 1 & 0 \\ 0 & 1 \end{bmatrix}$.

▷ Reflection about the x-axis: $\begin{bmatrix} 1 & 0 \\ 0 & -1 \end{bmatrix}$.

▷ Reflection about the y-axis: $\begin{bmatrix} -1 & 0 \\ 0 & 1 \end{bmatrix}$.

[1]Homogeneous coordinates also play a fundamental role in the study of projective geometry.

▷ Reflection about the line $y = x$: $\begin{bmatrix} 0 & 1 \\ 1 & 0 \end{bmatrix}$.

▷ Rotation counterclockwise through a positive angle θ: $\begin{bmatrix} \cos(\theta) & -\sin(\theta) \\ \sin(\theta) & \cos(\theta) \end{bmatrix}$.

(In MATLAB angle θ must be in radians.)

▷ Scaling by h in the x-direction and by k in the y-direction: $\begin{bmatrix} h & 0 \\ 0 & k \end{bmatrix}$.

▷ Shear in the x-direction by the factor k: $\begin{bmatrix} 1 & k \\ 0 & 1 \end{bmatrix}$.

▷ Shear in the y-direction by the factor k: $\begin{bmatrix} 1 & 0 \\ k & 1 \end{bmatrix}$.

The translation of a point, vector, or object defined by a set of points in the plane, is performed by adding the same quantity Δx to each x-coordinate and the same quantity Δy to each y-coordinate. (We emphasize that Δx and Δy are not required to be equal in magnitude.) We illustrate this in Figure 2 for a point in R^2, where the coordinates of the translated point are:

$$(\boldsymbol{x}^*, \boldsymbol{y}^*) = (\boldsymbol{x} + \Delta \boldsymbol{x}, \boldsymbol{y} + \Delta \boldsymbol{y}).$$

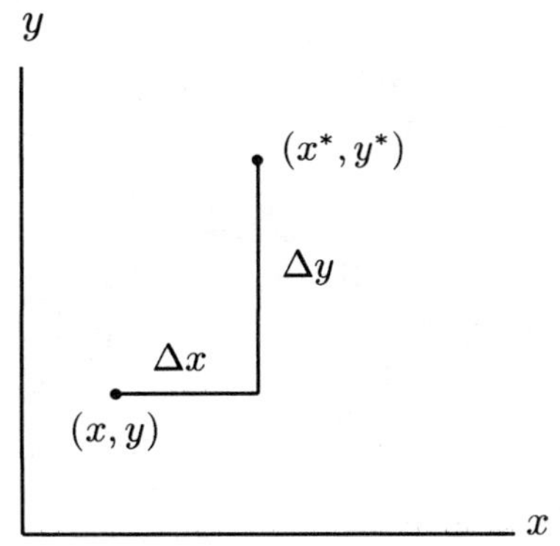

Figure 2

In order to have scalings, projections, and rotations "play together nicely" with translations we change the space in which we work. In order to employ matrix multiplication to perform translations and avoid the direct additions of changes to individual coordinates, we adjoin another component to vectors and border matrices (See Figure 2.) with another row and column. This change is said to use **homogeneous coordinates**. To use homogeneous coordinates we make the following identifications.

A vector $\begin{bmatrix} x \\ y \end{bmatrix}$ in R^2 in identified with the vector $\begin{bmatrix} x \\ y \\ 1 \end{bmatrix}$ in R^3.

The first two coordinates are the same and the third coordinate is 1.

Each of the matrices $\boldsymbol{A}$ associated with plane linear transformations is now identified with a 3×3 matrix of the form

$$\begin{bmatrix} \boldsymbol{A} & \begin{bmatrix} 0 \\ 0 \end{bmatrix} \\ \begin{bmatrix} 0 & 0 \end{bmatrix} & [1] \end{bmatrix} = \begin{bmatrix} a_{11} & a_{12} & 0 \\ a_{21} & a_{22} & 0 \\ 0 & 0 & 1 \end{bmatrix}$$

For example when using homogeneous coordinates for a reflection about the y-axis the corresponding matrix is the 3×3 matrix $\begin{bmatrix} -1 & 0 & 0 \\ 0 & 1 & 0 \\ 0 & 0 & 1 \end{bmatrix}$. Also when using homogeneous coordinates for a rotation by an angle θ the corresponding matrix is the 3×3 matrix $\begin{bmatrix} \cos(\theta) & -\sin(\theta) & 0 \\ \sin(\theta) & \cos(\theta) & 0 \\ 0 & 0 & 1 \end{bmatrix}$.

A translation can be performed by matrix multiplication on data expressed in homogeneous coordinates using the 3×3 matrix $\begin{bmatrix} 1 & 0 & \Delta x \\ 0 & 1 & \Delta y \\ 0 & 0 & 1 \end{bmatrix}$.

We have

$$\begin{bmatrix} 1 & 0 & \Delta x \\ 0 & 1 & \Delta y \\ 0 & 0 & 1 \end{bmatrix} \begin{bmatrix} x \\ y \\ 1 \end{bmatrix} = \begin{bmatrix} x + \Delta x \\ y + \Delta y \\ 1 \end{bmatrix}.$$

Experiments Employing Homogeneous Coordinates

Section 11.1 described plane linear transformations geometrically and related them algebraically to their matrix transformations. This section introduces a broader range of transformations with a view toward providing geometric motivation for the determinant of a matrix as introduced in Lab 8. Moreover, the present section uses a series of experiments to provide a physical meaning to the size of the determinant of a matrix used in a plane linear transformation and a physical setting to educe a formula for the determinant of the product of two matrices.

The main tool is the routine **matrixmaps**. Typing this command produces a screen with several components. The menu on the right-hand side lists seven objects: triangle, house, rectangle, arrow, square, semi-circle, and polygon. Begin by selecting the square. To view the object, click the View button. A menu appears for obtaining help on using five transformations: reflection, scaling, rotation, translation, and shears. Select translation. A pop-up window provides directions for entering a matrix for this transformation. Thus the matrix for translation by $a = 3$ units in the x-direction and $b = 2$ units in the y-direction is

$$\boldsymbol{A} = \begin{bmatrix} 1 & 0 & 3 \\ 0 & 1 & 2 \\ 0 & 0 & 1 \end{bmatrix}$$

Exit the window by clicking "OK". Next click the *MATRIX* button. This produces another menu of help options, but since we have already ascertained the proper form for a translation matrix $\boldsymbol{A}$ type 0 to QUIT the help and press ENTER to continue. Then enter the matrix $\boldsymbol{A}$ in the form [1 0 3; 0 1 2; 0 0 1] and press ENTER. Now, click on the MAP IT button to display the image of the object. Notice that the image of the original square (in red) is another square (in blue).

Now that we know how to use **matrixmaps** we use this routine to provide geometric motivation for the concept of the determinant of a matrix. Click *Restart*, select the square again, and then click on the *View*. To show an alternate way to proceed, click on the *MATRIX* button at once. We intend to scale the square so type 2 for directions for entering the appropriate matrix. Type 0 to QUIT the help and press ENTER to continue. We will use

$$\boldsymbol{A} = \begin{bmatrix} 2 & 0 & 0 \\ 0 & 3 & 0 \\ 0 & 0 & 1 \end{bmatrix}$$

for scaling by 2 units in the x-direction and 3 units in the y-direction, so enter this matrix using the form of a scaling matrix, and then press ENTER. Click the MAP IT button and notice that the image of the square is a rectangle. For help in determining the areas of the respective figures, click the Grid On button from the menu in the bottom right-hand part of the screen. Notice that the area of the image is six times the area of the object. (Explain why this is true.)

Let us investigate what happens when a different object is scaled in the same way. Click *Restart* and select the rectangle. By clicking on *View* and on *Grid On* you can see that the area of the 4×3 object is 12. Click on the *MATRIX* button, type 0 to QUIT the help, press ENTER to continue, enter the same scaling matrix (use the up-arrow key several times to recall the previous scaling matrix, if desired), and press ENTER again. Click the MAP IT button and notice that the image is an 8×9 rectangle. Thus once again the area of the image (72) is 6 times the area of the object (12).

What is the significance of the number 6? Notice that $\det(\boldsymbol{A})=6$. This suggests that the determinant of $\boldsymbol{A}$ provides a **magnification factor** for this scaling. Use **matrixmaps** to verify this result on a different figure, the House. By turning on the grid you can see that the area of the object is 5. (Explain why.) Consequently the area of its image after scaling by 2 units in the x-direction and 3 units in the y-direction should be 30. This conclusion can be checked by observing that the image consists of six 2×2 squares and two 2×3 right triangles. We display the general property that these experiments suggest

$$\text{Area(image)} = |\det(\boldsymbol{A})| * \text{Area(object)}$$

(Since we can scale by negative values, it is possible that $\det(\boldsymbol{A}) < 0$, hence because of physical considerations we have included absolute values around $\det(\boldsymbol{A})$.)

It is also possible to verify this property for an object that is not a polygon. Restart **matrixmaps**, select the semi-circle, and turn the Grid On to see that its radius is 1; hence its area is $\pi/2$. Look closely at its image after the same scaling. It is a half of an ellipse with semi-minor axis of length $a = 2$ and semi-major axis of length $b = 3$. Recall that the area of a half of an ellipse with semi-minor axis of length a and semi-major axis of length b is $\pi ab/2$. Hence the area of the image is $6\pi/2$, verifying the stated property once again.

Next we provide a geometric justification for the algebraic fact that

$$\det(\boldsymbol{AB}) = \det(\boldsymbol{A})\det(\boldsymbol{B})$$

by examining the relationship between the area of the object (semi-circle) and the area of a composite image. We have just viewed the image, a half of an ellipse with area $6\pi/2$. Now click the Composite button to form the composition of transformations. This time we choose a scaling 4 units in the x-direction and 2 units in the y-direction, so click the *MATRIX* button and enter the matrix $\boldsymbol{B}$ in the format $[4\ 0\ 0; 0\ 2\ 0; 0\ 0\ 1]$. Notice that $\det(\boldsymbol{B}) = 8$. Clicking the MAP IT button reveals the image (in blue) and the composite (in magenta). The composite image is half of an ellipse with semi-major axis of length 8 (because it runs from 8 on the left to 24 on the right) and semi-minor axis of length 6 (because it is 12-6). Therefore the area of the composite image is $48\pi/2$, which is 48 times the area of the object ($\pi/2$). Since $48 = 6 \times 8$, this result confirms geometrically that $\det(\boldsymbol{AB}) = \det(\boldsymbol{A})\det(\boldsymbol{B})$.

To confirm this result for the same transformations on a different figure, restart with the House and then apply scaling by 2 units in the x-direction and 3 units in the y-direction. We have seen that the area of the object is 5 and the area of the image is 30. Now compose this transformation with scaling by 4 units in the x-direction and 2 units in the y-direction. The blue image lies inside the larger magenta composite image, which is formed from a 16×12 rectangle and two 8×6 right triangles, so the area of the composite image is 240, which is 48 times the area of the object. This provides a second geometric confirmation that $\det(\boldsymbol{AB}) = \det(\boldsymbol{A})\det(\boldsymbol{B})$. Exit **matrixmaps** by pressing the Quit button.

Exercises 11.2

1. Use **matrixmaps** to perform each of the following.

a) Select the House object. Determine a matrix $\boldsymbol{A}$ (in homogeneous coordinate form) so that the image is a house only half as wide.

$\boldsymbol{A} =$

b) Select the House object. Determine a matrix $\boldsymbol{A}$ (in homogeneous coordinate form) so that the image is a house only half as wide and half as tall.

$\boldsymbol{A} =$

c) Select the House object. Use $\boldsymbol{A} = \begin{bmatrix} 1 & 1 & 0 \\ 1 & 1 & 0 \\ 0 & 0 & 1 \end{bmatrix}$, determine the image and explain why the house is uninhabitable.

d) Select the House object. What matrix $\boldsymbol{A}$ will 'shrink' the house to a single point? Verify your choice.

$\boldsymbol{A} =$

2. Use **matrixmaps** to perform each of the following.

a) Select the Arrow object. Determine a matrix $\boldsymbol{A}$ so that the image is an arrow pointed in the opposite direction.

$\boldsymbol{A} =$

b) Select the Arrow object. Determine a matrix $\boldsymbol{A}$ so that the image is an arrow pointed in the same direction but only half as long.

$\boldsymbol{A} =$

c) Select the Arrow object and use the matrix $\boldsymbol{A} = \begin{bmatrix} \cos(\pi/4) & -\sin(\pi/4 & 0 \\ \sin(\pi/4) & \cos(\pi/4) & 0 \\ 0 & 0 & 1 \end{bmatrix}$.
Describe the resulting image. What angle does it make with the positive horizontal

axis? To help answer this question, use the Grid On button and then inspect the grid generated on the mapped arrow. (Note: $\pi/4$ is entered as pi/4 in MATLAB .)

d) Using part c determine the coordinates of the top end of the arrow.

3. Use **matrixmaps** to perform each of the following using matrices

$$\boldsymbol{A} = \begin{bmatrix} 3 & 1 & 0 \\ 0 & 2 & 0 \\ 0 & 0 & 1 \end{bmatrix} \text{ and } \boldsymbol{B} = \begin{bmatrix} -1 & 2 & 0 \\ 2 & 0 & 0 \\ 0 & 0 & 1 \end{bmatrix}.$$

a) Choose the rectangle object. Use matrix $\boldsymbol{A}$ to map the rectangle then form the composite map by applying matrix $\boldsymbol{B}$ to this image. Carefully make a sketch of the final image or print the screen display.

b) Restart and again choose the rectangle object. Use matrix $\boldsymbol{B}$ to map the rectangle then form the composite map by applying matrix $\boldsymbol{A}$ to this image. Carefully make a sketch of the final image or print the screen display.

c) Are the composite images in parts a) and b) the same? Explain why or why not using an algebraic argument.

4. Use **matrixmaps** to perform each of the following.

a) Select the square object. Determine a matrix $\boldsymbol{A}$ (in homogeneous coordinates form) so that the upper right corner of the object appears at the point $(2, 3)$.

$\boldsymbol{A} =$

b) Select the square object. Determine a matrix $\boldsymbol{A}$ (in homogeneous coordinates form) so that the upper left corner of the object appears at the point $(2, 3)$.

$\boldsymbol{A} =$

c) Select the square object. Determine a matrix $\boldsymbol{A}$ (in homogeneous coordinates form) so that the image of the square is the following image.

$\boldsymbol{A} =$

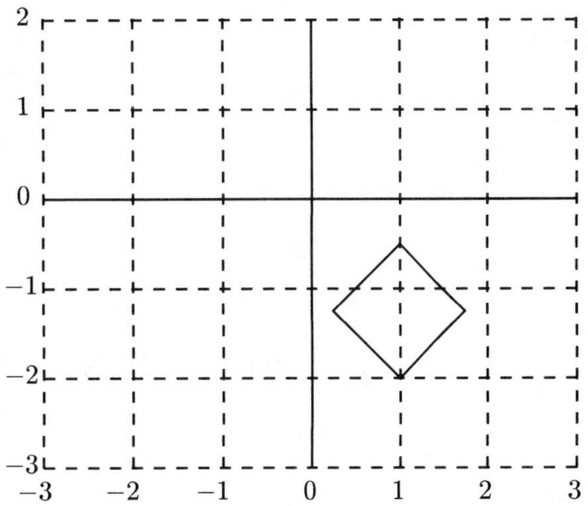

d) Select the square object. Determine a matrix $\boldsymbol{A}$ (in homogeneous coordinates form) so that the image of the square is the following image.

$A =$

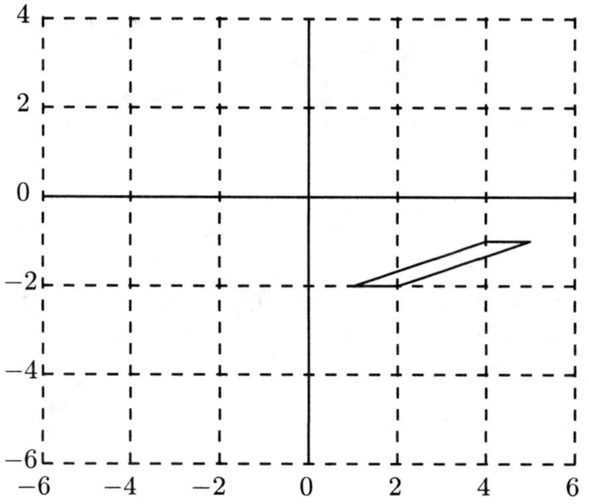

<< NOTES; COMMENTS; IDEAS >>

LAB 12

Linear Transformations

Topics: image; matrix transformation; matrix representation; coordinates of a vector relative to a basis; range, column space; pre-image; kernel, null space.

Introduction

We have seen many of MATLAB 's numerical, graphical, and symbolic capabilities. This lab continues this approach for linear transformations from R^n to R^m. Section 12.1 introduces linear transformations from R^n to R^m, while Section 12.2 discusses matrix representations. Sections 12.3 and 12.4 show how to compute the range and kernel in terms of two fundamental subspaces associated with the matrix representation, the column space and the null space, respectively.

This lab does not require a knowledge of Lab 11. However, it relies extensively on Sections 2.1, 5.1, 6.1 to 6.4, and 7.2.

Section 12.1

Definition and Properties

A linear transformation T from R^n to R^m is a function which satisfies the two properties

$$\mathrm{T}(\boldsymbol{u}+\boldsymbol{v}) = \mathrm{T}(\boldsymbol{u}) + \mathrm{T}(\boldsymbol{v}), \text{ for any vectors } \boldsymbol{u} \text{ and } \boldsymbol{v} \text{ in } \mathrm{R}^n$$
$$\mathrm{T}(c{*}\boldsymbol{u}) = c{*}\mathrm{T}(\boldsymbol{u}), \text{ for any real scalar } c \text{ and any vector } \boldsymbol{u} \text{ in } \mathrm{R}^n$$

For $\boldsymbol{u}$ in R^n, we call $\mathrm{T}(\boldsymbol{u})$ the **image** of vector $\boldsymbol{u}$. Note that the image $\mathrm{T}(\boldsymbol{u})$ is in R^m.

The two properties above can be combined into the single expression

$$\mathrm{T}(c{*}\boldsymbol{u}+k{*}\boldsymbol{v}) = c{*}\mathrm{T}(\boldsymbol{u})+k{*}\mathrm{T}(\boldsymbol{v})$$

T is a linear transformation if and only if the preceding expression is valid for all scalars c and k and all vectors $\boldsymbol{u}$ and $\boldsymbol{v}$ in R^n. This alternative is interpreted to say that a linear transformation 'splits apart' linear combinations in a natural way. Of course, if T is linear then this 'splitting' applies to linear combinations of any number of vectors. *If T is a linear transformation then the image of a linear combination is the corresponding linear combination of the images (of the individual vectors).*

An important linear transformation is the function defined by a matrix multiply. If $\boldsymbol{M}$ is an $m \times n$ matrix then for $\boldsymbol{u}$ in R^n the product $\boldsymbol{M} * \boldsymbol{u}$ defines a function from R^n to R^m;

$$\mathrm{T}(\boldsymbol{u}) = \boldsymbol{M} * \boldsymbol{u}$$

Show that $\mathrm{T}(\boldsymbol{u}) = \boldsymbol{M} * \boldsymbol{u}$ is a linear transformation from R^n to R^m. (Hint: use properties of matrix multiplication.) Record your work below.

The linear transformation defined by the multiplication of a vector by a matrix appears in many situations. We adopt the terminology *'matrix transformation'* to emphasize its importance. In particular if the definition of a transformation is given by an expression and we observe that the images can be computed by a matrix transformation, then we can immediately infer that T is a linear transformation. For example, suppose that transformation T is given by

$$\mathrm{T}\left(\begin{bmatrix} x \\ y \end{bmatrix}\right) = \begin{bmatrix} x \\ 0 \end{bmatrix}$$

T defines a function from R^2 to R^2 such that image of vector $\boldsymbol{u} = \begin{bmatrix} x \\ y \end{bmatrix}$ is $\begin{bmatrix} x \\ 0 \end{bmatrix}$. Geometrically we have

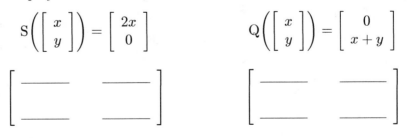

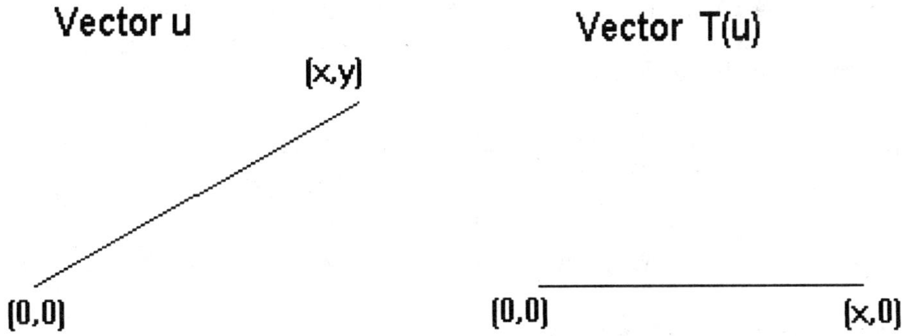

To show that T is a linear transformation you could proceed algebraically and verify the two properties in the definition. Or you could determine a 2×2 matrix $\boldsymbol{M}$ such that $\mathrm{T}(\boldsymbol{u}) = \boldsymbol{M} * \boldsymbol{u}$.

Find $\boldsymbol{M}$ such that $\mathrm{T}(\boldsymbol{u}) = \boldsymbol{M} * \boldsymbol{u}$. Record matrix $\boldsymbol{M}$ below.

$$\boldsymbol{M} = \begin{bmatrix} \_\_\_\_ & \_\_\_\_ \\ \_\_\_\_ & \_\_\_\_ \end{bmatrix}$$

Warning: just because you are not able to find a matrix $\boldsymbol{M}$ that works does not imply the transformation T is not linear. Possibly you were just not clever enough to figure out a matrix that works.

Show that each of the following transformations is linear by verifying that it is a matrix transformation. Display the matrix in each case.

$$\mathrm{S}\left(\begin{bmatrix} x \\ y \end{bmatrix}\right) = \begin{bmatrix} 2x \\ 0 \end{bmatrix} \qquad \mathrm{Q}\left(\begin{bmatrix} x \\ y \end{bmatrix}\right) = \begin{bmatrix} 0 \\ x + y \end{bmatrix}$$

$$\begin{bmatrix} \_\_\_\_ & \_\_\_\_ \\ \_\_\_\_ & \_\_\_\_ \end{bmatrix} \qquad \begin{bmatrix} \_\_\_\_ & \_\_\_\_ \\ \_\_\_\_ & \_\_\_\_ \end{bmatrix}$$

$$\mathrm{R}\left(\begin{bmatrix} x \\ y \end{bmatrix}\right) = \begin{bmatrix} 2x - y \\ y \end{bmatrix} \qquad \mathrm{Z}\left(\begin{bmatrix} x \\ y \end{bmatrix}\right) = \begin{bmatrix} 0 \\ 0 \end{bmatrix}$$

$$\begin{bmatrix} \underline{\qquad} & \underline{\qquad} \\ \underline{\qquad} & \underline{\qquad} \end{bmatrix} \qquad \begin{bmatrix} \underline{\qquad} & \underline{\qquad} \\ \underline{\qquad} & \underline{\qquad} \end{bmatrix}$$

Check your conjecture for the matrix transformation using matrix multiplication.

There is certainly a close relationship between linear transformations and linear combinations as we indicated above. But the connection can be extended further to determine which transformations from R^n to R^m are linear. From the preceding experiments you may have noticed a pattern for determining the matrix. In particular, since we were to have a 'matrix transformation', $\boldsymbol{M} * \boldsymbol{u}$, the entries of the image are linear combinations of the entries of vector $\boldsymbol{u}$. From matrix multiplication we have that

$$\boldsymbol{M} * \boldsymbol{u} = \begin{bmatrix} row_1(\boldsymbol{M}) * \boldsymbol{u} \\ row_2(\boldsymbol{M}) * \boldsymbol{u} \\ \vdots \\ row_m(\boldsymbol{M}) * \boldsymbol{u} \end{bmatrix} = \begin{bmatrix} m_{11}u_1 + m_{12}u_2 + \cdots + m_{1n}u_n \\ m_{21}u_1 + m_{22}u_2 + \cdots + m_{2n}u_n \\ \vdots \\ m_{m1}u_1 + m_{m2}u_2 + \cdots + m_{mn}u_n \end{bmatrix}$$

We use this observation in two ways:

1. If the entries of the image $\mathrm{T}(\boldsymbol{u})$ are not linear combinations of the entries of vector $\boldsymbol{u}$, then T is not a matrix transformation.

2. If the entries of the image $\mathrm{T}(\boldsymbol{u})$ are linear combinations of the entries of vector $\boldsymbol{u}$, then the matrix for the transformation has its rows formed from the coefficients used in the linear combinations.

To complete the connections started above, it would be convenient to conclude that *a transformation T from R^n to R^m is linear if and only if T is a matrix transformation.* Currently we can not draw this conclusion, but we deal with this issue in the next section.

For each of the following transformations determine if they are linear or not. If it is linear display the matrix of the transformation. If it is not linear explicitly show that one of the properties for a linear transformation is violated. Put your responses near the definition of the transformation.

$$\mathrm{R}\left(\begin{bmatrix} x \\ y \\ z \end{bmatrix}\right) = \begin{bmatrix} 2x - y \\ xy \\ 0 \end{bmatrix} \qquad \mathrm{S}\left(\begin{bmatrix} x \\ y \end{bmatrix}\right) = \begin{bmatrix} x^2 \\ 4x + 2y \end{bmatrix}$$

$$R\left(\begin{bmatrix} x \\ y \\ z \end{bmatrix}\right) = \begin{bmatrix} 5y+z \\ x-6y+2z \\ 0 \end{bmatrix} \qquad S\left(\begin{bmatrix} x \\ y \end{bmatrix}\right) = \begin{bmatrix} x+8 \\ -3x+9y \end{bmatrix}$$

$$R\left(\begin{bmatrix} x \\ y \\ z \end{bmatrix}\right) = \begin{bmatrix} 5y+z \\ x-6y+2z \\ 5 \end{bmatrix} \qquad S\left(\begin{bmatrix} x \\ y \end{bmatrix}\right) = \begin{bmatrix} 1/x \\ -3x+9y \end{bmatrix}$$

Since every vector in R^n is a linear combination of vectors in any basis for R^n, knowing the action of the linear transformation on the basis vectors gives us enough information to determine the image of any vector. This follows from the properties of linear combinations and the behavior of a linear transformation. Briefly, if we know that

$$\{\boldsymbol{w_1}, \boldsymbol{w_2}, \ldots, \boldsymbol{w_n}\}$$

is a basis for R^n and linear transformation T gives the images of these basis vectors as

$$\{\mathrm{T}(\boldsymbol{w_1})=\boldsymbol{p_1},\ \mathrm{T}(\boldsymbol{w_2})=\boldsymbol{p_2},\ \ldots,\ \mathrm{T}(\boldsymbol{w_n})=\boldsymbol{p_n}\}$$

then

$$\mathrm{T}(\boldsymbol{u}) = c_1\boldsymbol{p_1}+c_2\boldsymbol{p_2}+\ldots+c_n\boldsymbol{p_n}$$

where the coefficients $\{c_1, c_2, \ldots, c_n\}$ are those scalars that are used to express $\boldsymbol{u}$ in terms of the basis $\{\boldsymbol{w_1}, \boldsymbol{w_2}, \ldots, \boldsymbol{w_n}\}$;

$$\boldsymbol{u} = c_1\boldsymbol{w_1}+c_2\boldsymbol{w_2}+\ldots+c_n\boldsymbol{w_n}$$

Example 1. Let $B = \{\boldsymbol{w_1}, \boldsymbol{w_2}, \boldsymbol{w_3}\}$ be a basis of R^3, where

$$\boldsymbol{w_1} = \begin{bmatrix} 1 \\ 0 \\ 0 \end{bmatrix}, \qquad \boldsymbol{w_2} = \begin{bmatrix} 1 \\ 1 \\ 0 \end{bmatrix}, \qquad \boldsymbol{w_3} = \begin{bmatrix} 1 \\ 1 \\ 1 \end{bmatrix}$$

Let T be a linear transformation from R^3 to R^3 for which

$$T(\boldsymbol{w_1}) = \begin{bmatrix} 1 \\ 2 \\ 3 \end{bmatrix}, \qquad T(\boldsymbol{w_2}) = \begin{bmatrix} -1 \\ 0 \\ 4 \end{bmatrix}, \qquad T(\boldsymbol{w_3}) = \begin{bmatrix} 0 \\ 2 \\ -1 \end{bmatrix}$$

To compute $T(\boldsymbol{u})$ for $\boldsymbol{u}= \begin{bmatrix} 11 \\ 13 \\ 8 \end{bmatrix}$ using the analysis above, we need to determine the coefficients c_1, c_2, and c_3 so that

$$\boldsymbol{u}= c_1\boldsymbol{w_1}+c_2\boldsymbol{w_2} +c_3\boldsymbol{w_3}$$

and then form the corresponding linear combination of the images of the basis vectors.

Find a linear system whose solution gives the coefficients c_1, c_2, and c_3. Display the coefficient matrix, right-hand side, and the solution below. (Use MATLAB to compute the solution.)

You should find $c_1 = -2, c_2 = 5$, and $c_3 = 8$.

Determine $T(\boldsymbol{u})$. Show your steps below.

We have

$$T(\boldsymbol{u})= c_1T(\boldsymbol{w_1})+c_2T(\boldsymbol{w_2}) +c_3T(\boldsymbol{w_3})$$

Express this linear combination of columns as a matrix multiply; that is, find a matrix $\boldsymbol{M}$ such that $T(\boldsymbol{u})= \boldsymbol{M}*$ *'some vector'*. Display $\boldsymbol{M}$ and the *'vector'* in the space below.

Explain the connection between $\boldsymbol{M}$ and the *'vector'*.

Exercises 12.1

1. For each of the following transformations determine if it is linear or not. If it is linear display the matrix of the transformation. If it is not linear explicitly show that one of the properties for a linear transformation is violated.

a) $T\left(\begin{bmatrix} x \\ y \end{bmatrix}\right) = \begin{bmatrix} -x \\ -y \end{bmatrix}$

b) $T\left(\begin{bmatrix} x \\ y \\ z \end{bmatrix}\right) = \begin{bmatrix} x+y \\ z-1 \end{bmatrix}$

c) $T\left(\begin{bmatrix} x \\ y \end{bmatrix}\right) = \begin{bmatrix} 2x-y \\ x-3y \\ 4y \end{bmatrix}$

2. The transformation T from R^4 to R, the real numbers, is defined as follows:

$$T\left(\begin{bmatrix} x \\ y \\ z \\ w \end{bmatrix}\right) = x+y+z+w$$

If T is a linear transformation display the matrix of the transformation. If T is not linear explicitly show that one of the properties for a linear transformation is violated.

Express transformation T in terms of a dot product. Explain how this alternate formulation can be used to reveal whether T is linear or not.

3. Let B $= \{\boldsymbol{w_1}, \boldsymbol{w_2}, \boldsymbol{w_3}\}$ be the basis of R^3 with $\boldsymbol{w_1} = \begin{bmatrix} 1 \\ 1 \\ 0 \end{bmatrix}$, $\boldsymbol{w_2} = \begin{bmatrix} 2 \\ 0 \\ 1 \end{bmatrix}$, $\boldsymbol{w_3} = \begin{bmatrix} 0 \\ 1 \\ 2 \end{bmatrix}$.
Let T be a linear transformation such that the images of the basis vectors are

$$T(\boldsymbol{w_1}) = \begin{bmatrix} 1 \\ 0 \\ -2 \end{bmatrix}, T(\boldsymbol{w_2}) = \begin{bmatrix} -1 \\ 0 \\ 1 \end{bmatrix}, T(\boldsymbol{w_3}) = \begin{bmatrix} 0 \\ 3 \\ -5 \end{bmatrix}$$

a) Compute $T(\boldsymbol{u})$ for $\boldsymbol{u} = \begin{bmatrix} 1 \\ 1 \\ -5 \end{bmatrix}$.

b) Determine the matrix of the linear transformation T.

4. Let B $= \{\boldsymbol{w_1}, \boldsymbol{w_2}, \boldsymbol{w_3}, \boldsymbol{w_4}\}$ be the basis of R^4 with $\boldsymbol{w_1} = \begin{bmatrix} 1 \\ 0 \\ 0 \\ 0 \end{bmatrix}$, $\boldsymbol{w_2} = \begin{bmatrix} 0 \\ 1 \\ 1 \\ 0 \end{bmatrix}$, $\boldsymbol{w_3} = \begin{bmatrix} 0 \\ 0 \\ 2 \\ 1 \end{bmatrix}$
$\boldsymbol{w_4} = \begin{bmatrix} 0 \\ 1 \\ 0 \\ 1 \end{bmatrix}$. Let T be a linear transformation such that the images of the basis vectors are

$$T(\boldsymbol{w_1}) = \begin{bmatrix} 1 \\ 0 \\ -2 \\ 3 \end{bmatrix}, T(\boldsymbol{w_2}) = \begin{bmatrix} -1 \\ 0 \\ 1 \\ 0 \end{bmatrix}, T(\boldsymbol{w_3}) = \begin{bmatrix} 0 \\ 3 \\ -5 \\ 1 \end{bmatrix}, T(\boldsymbol{w_3}) = \begin{bmatrix} -6 \\ 3 \\ 5 \\ 7 \end{bmatrix}$$

a) Compute $T(\boldsymbol{u})$ for $\boldsymbol{u} = \begin{bmatrix} 1 \\ -2 \\ 2 \\ 1 \end{bmatrix}$.

b) Compute T($\boldsymbol{u}$) for $\boldsymbol{u} = \begin{bmatrix} 0 \\ 0 \\ 0 \\ 1 \end{bmatrix}$.

c) Compute T($\boldsymbol{u}$) for $\boldsymbol{u} = \begin{bmatrix} 0 \\ 0 \\ 0 \\ 0 \end{bmatrix}$.

(Do you really have to do any computation here? Explain.)

d) Determine the matrix of linear transformation T.

5. For any linear transformation T, not just matrix transformations, explain why the image of the zero vector must be the zero vector.

6. Explain how to use the result in the preceding exercise to develop a test to show a transformation is not linear.

Section 12.2

Matrix Representations

A fundamental result of linear algebra is that, loosely speaking, 'every linear transformation is a matrix multiplication.' In the previous section we showed that if a transformation T from R^n to R^m is a *matrix transformation* then T is linear. We also showed that if the entries of the image $T(\boldsymbol{u})$ are linear combinations of the entries of $\boldsymbol{u}$ then T is a matrix transformation (and hence linear). Here we show that every linear transformation from R^n to R^m is a matrix transformation and find its matrix representation. We also explore matrix representations of a linear transformation relative to different bases.

Let $\{\boldsymbol{w_1}, \boldsymbol{w_2}, \ldots, \boldsymbol{w_n}\}$ be a basis for R^n and T be a linear transformation from R^n to R^m such that the images of the basis vectors are

$$\{T(\boldsymbol{w_1})=\boldsymbol{p_1},\ T(\boldsymbol{w_2})=\boldsymbol{p_2},\ \ldots,\ T(\boldsymbol{w_n})=\boldsymbol{p_n}\}$$

For an arbitrary vector $\boldsymbol{u}$ in R^n there exist scalars $c_1, c_2, ..., c_n$ such that

$$\boldsymbol{u} = c_1\boldsymbol{w_1}+c_2\boldsymbol{w_2}+\cdots+c_n\boldsymbol{w_n}$$

and since T is linear

$$T(\boldsymbol{u}) = c_1\boldsymbol{p_1}+c_2\boldsymbol{p_2}+\ldots+c_n\boldsymbol{p_n}$$

Hence image $T(\boldsymbol{u})$ is a linear combination of columns $\boldsymbol{p_i}$ in R^m which implies that $T(\boldsymbol{u})$ can be represented as a matrix multiply. We have

$$T(\boldsymbol{u}) = [\boldsymbol{p_1}\ \boldsymbol{p_2}\ \cdots\ \boldsymbol{p_n}] \begin{bmatrix} c_1 \\ c_2 \\ \vdots \\ c_n \end{bmatrix} = \boldsymbol{Mc}$$

where $\boldsymbol{M}$ is an $m \times n$ matrix and $\mathbf{c}$ is a vector in R^n:

$$\boldsymbol{M} = [\boldsymbol{p_1}\ \boldsymbol{p_2}\ \cdots\ \boldsymbol{p_n}], \qquad \mathbf{c} = \begin{bmatrix} c_1 \\ c_2 \\ \vdots \\ c_n \end{bmatrix}$$

The columns of $\boldsymbol{M}$ are the images of the basis from R^n and $\boldsymbol{M}$ is called the **matrix representation** of linear transformation T. Thus T is a matrix transformation. Combining this with the results in Section 12.1, we have the following important result. [1]

[1]This result can be generalized to linear transformations between any two vector spaces.

A transformation T from R^n to R^m is linear if and only if it is a matrix transformation.

The expression for the matrix representation $\boldsymbol{M}$ of linear transformation T depends upon the basis used in R^n. Consider, for instance, the plane linear transformation T from R^3 to R^2 defined geometrically as the projection onto the xy-plane: $T\left(\begin{bmatrix} x \\ y \\ z \end{bmatrix}\right) = \begin{bmatrix} x \\ y \end{bmatrix}$. For the standard basis $\boldsymbol{e_1} = \begin{bmatrix} 1 \\ 0 \\ 0 \end{bmatrix}$, $\boldsymbol{e_2} = \begin{bmatrix} 0 \\ 1 \\ 0 \end{bmatrix}$, and $\boldsymbol{e_3} = \begin{bmatrix} 0 \\ 0 \\ 1 \end{bmatrix}$,

$$\boldsymbol{T}(\boldsymbol{e_1}) = \begin{bmatrix} 1 \\ 0 \end{bmatrix}, \quad \boldsymbol{T}(\boldsymbol{e_2}) = \begin{bmatrix} 0 \\ 1 \end{bmatrix}, \quad \boldsymbol{T}(\boldsymbol{e_3}) = \begin{bmatrix} 0 \\ 0 \end{bmatrix}$$

Thus the matrix representation of T relative to the standard basis in R^3 is

$$\boldsymbol{M} = \begin{bmatrix} \boldsymbol{T}(\boldsymbol{e_1}) & \boldsymbol{T}(\boldsymbol{e_2}) & \boldsymbol{T}(\boldsymbol{e_3}) \end{bmatrix} = \begin{bmatrix} 1 & 0 & 0 \\ 0 & 1 & 0 \end{bmatrix}$$

and we can compute images of T as

$$T\left(\begin{bmatrix} x \\ y \\ z \end{bmatrix}\right) = \boldsymbol{M}\begin{bmatrix} x \\ y \\ z \end{bmatrix} = \begin{bmatrix} 1 & 0 & 0 \\ 0 & 1 & 0 \end{bmatrix}\begin{bmatrix} x \\ y \\ z \end{bmatrix} = \begin{bmatrix} x \\ y \end{bmatrix}$$

In certain situations it is advantageous to have vectors expressed in terms of a basis that is 'convenient' for the situation at hand. For instance suppose that the motion of an object in R^3 is restricted to a plane and we need to project its path onto the xy-plane. The *coordinates* of the points along the object's path are given in terms of a convenient basis for the plane. It follows that the matrix representation of the projection in this case is formed from the (projection) images of the two basis vectors for the plane. In the xy-plane the coordinates of the image of a point along the object's path are found from the product of the matrix representation and the coordinates of points along the path in the plane. For example, suppose that the object's path lies in plane

$$x - 2y + 4z = 0$$

and we use basis $\boldsymbol{w_1} = \begin{bmatrix} 0 \\ 2 \\ 1 \end{bmatrix}$ and $\boldsymbol{w_2} = \begin{bmatrix} 10 \\ 3 \\ -1 \end{bmatrix}$ for this plane. Then every point on the path is a linear combination of $\boldsymbol{w_1}$ and $\boldsymbol{w_2}$. We express the coordinates of such a point as the scalars used on the basis vectors to produce the point. (See Lab 7.) Suppose that point $\boldsymbol{v} = \begin{bmatrix} -4 \\ 0 \\ 1 \end{bmatrix}$ is on the object's path. Then the coordinates of this point in the plane relative to the basis $S = \{\boldsymbol{w_1}, \boldsymbol{w_2}\}$ are found by solving the linear system obtained from

$$c_1\boldsymbol{w_1} + c_2\boldsymbol{w_2} = \begin{bmatrix} -4 \\ 0 \\ 1 \end{bmatrix} \Longrightarrow \begin{bmatrix} 0 & 10 \\ 2 & 3 \\ 1 & -1 \end{bmatrix} \begin{bmatrix} c_1 \\ c_2 \end{bmatrix} = \begin{bmatrix} -4 \\ 0 \\ 1 \end{bmatrix}$$

In MATLAB we use command **c = [0 2 1;10 3 -1]'\[-4 0 1]'** to find that $\mathbf{c} = \begin{bmatrix} 0.6000 \\ -0.4000 \end{bmatrix}$. In coordinate notation as established in Lab 7 we have $[\boldsymbol{v}]_S = \begin{bmatrix} 0.6000 \\ -0.4000 \end{bmatrix}$. It follows that the projection of $\boldsymbol{v}$ into the xy-plane is

$$\boldsymbol{Mc} = [\mathrm{T}(\boldsymbol{w_1})\ \mathrm{T}(\boldsymbol{w_2})]\ [\boldsymbol{v}]_S = \begin{bmatrix} 0 & 10 \\ 2 & 3 \end{bmatrix} \begin{bmatrix} 0.6000 \\ -0.4000 \end{bmatrix} = \begin{bmatrix} -4 \\ 0 \end{bmatrix}$$

Of course this result can be obtained directly since we knew the coordinates of point $\boldsymbol{v}$ in the standard xyz-coordinate system. However, the value of the procedure is that we need only know the coordinates of points along the path in terms of the basis vectors $\boldsymbol{w_1}$ and $\boldsymbol{w_2}$. To illustrate this, let Q $= \{q_i,\ i = 1, 2, \ldots, 20\}$ be a set of points along the object's path with coordinates $\begin{bmatrix} t_i \\ t_i^2 - 1 \end{bmatrix}$ where $t_i = .1, .2, \ldots, 2$ relative to basis $\{\boldsymbol{w}_1,\ \boldsymbol{w}_2\}$. The projections of the points in set Q are obtained as $\boldsymbol{M}\begin{bmatrix} t_i \\ t_i^2 - 1 \end{bmatrix}$. (See Figure 1.) It is important to note that the coordinates of the images are obtained from the product of the matrix representation and the coordinates of the points in terms of the basis vectors $\{\boldsymbol{w}_1,\ \boldsymbol{w}_2\}$.

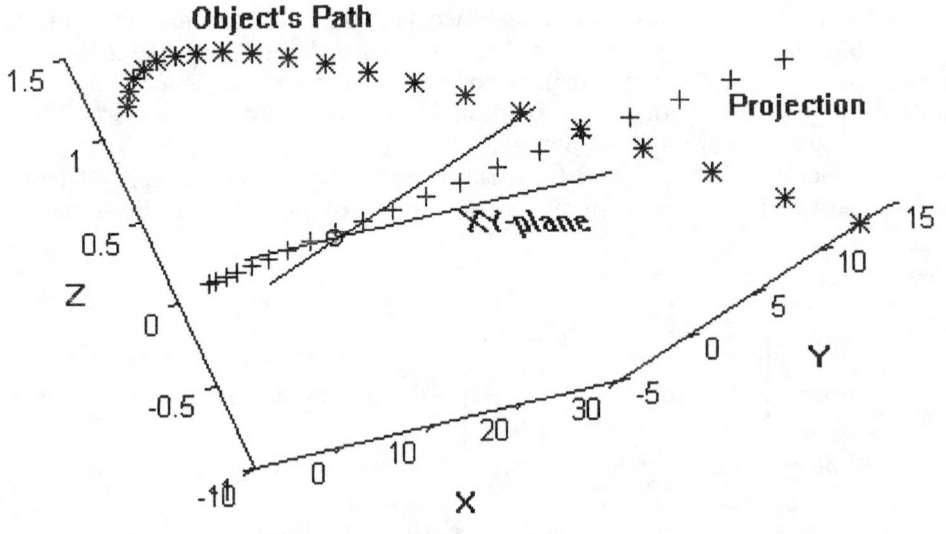

Figure 1.

There is an implicit assumption that the basis used for representing images in R^m is the standard or natural basis (the columns of the $m \times m$ identity matrix). If images in R^m are to be represented as linear combinations of a basis other than the standard basis, the matrix representation will change. In order to develop the matrix representation of T in such a case we briefly review some preliminary ideas.

Let T be a linear transformation from R^n to R^m, $\mathcal{A} = \{\boldsymbol{a_1}, \boldsymbol{a_2}, \ldots, \boldsymbol{a_n}\}$ be a basis for R^n, $\mathcal{B} = \{\boldsymbol{b_1}, \boldsymbol{b_2}, \ldots, \boldsymbol{b_m}\}$ be a basis for R^m, and $\boldsymbol{M}$ be the matrix representation of T relative to $\mathcal{A}$ and $\mathcal{B}$. Then the j-th column of $\boldsymbol{M}$ is the coordinate vector of the image $\mathrm{T}(\boldsymbol{a_j})$ relative to basis $\mathcal{B}$. Example 1 contains the essential ingredients for using MATLAB to determine the matrix representation of T relative to $\mathcal{A}$ and $\mathcal{B}$. The general approach is developed after the example.

<u>Example 1.</u> Let T be reflection about the y-axis in the plane. That is, T is the linear transformation from R^2 to R^2 such that $\mathrm{T}\left(\begin{bmatrix} x \\ y \end{bmatrix}\right) = \begin{bmatrix} -x \\ y \end{bmatrix}$. Find the matrix representation $\boldsymbol{M}$ of T relative to the bases $\mathcal{A} = \{\boldsymbol{a1}, \boldsymbol{a2}\}$ and $\mathcal{B} = \{\boldsymbol{b1}, \boldsymbol{b2}\}$, with

$$\boldsymbol{a1} = \begin{bmatrix} 5 \\ 2 \end{bmatrix}, \qquad \boldsymbol{a2} = \begin{bmatrix} 3 \\ 1 \end{bmatrix}, \qquad \boldsymbol{b1} = \begin{bmatrix} -1 \\ 2 \end{bmatrix}, \qquad \boldsymbol{b2} = \begin{bmatrix} 2 \\ 0 \end{bmatrix}$$

Enter the vectors $\boldsymbol{a1}, \boldsymbol{a2}, \boldsymbol{b1}, \boldsymbol{b2}$ into MATLAB . Define matrix $\boldsymbol{A}$= $[\boldsymbol{a1}, \boldsymbol{a2}]$ and $\boldsymbol{B}$= $[\boldsymbol{b1}, \boldsymbol{b2}]$. The first column $\boldsymbol{m1}$ of $\boldsymbol{M}$ is the coordinate vector of $\mathrm{T}(\boldsymbol{a1})$ relative to $\mathcal{B}$.

$$\mathrm{T}(\boldsymbol{a1}) = \begin{bmatrix} -5 \\ 2 \end{bmatrix} \qquad \boldsymbol{m1} = [\mathrm{T}(\boldsymbol{a1})]_{\mathcal{B}}$$

Then the coordinate vector of $\mathrm{T}(\boldsymbol{a1})$ relative to $\mathcal{B}$ is the set of coefficients that express $\mathrm{T}(\boldsymbol{a1})$ as a linear combination of $\boldsymbol{b1}$ and $\boldsymbol{b2}$. In MATLAB we have

m1 = B\[-5 2]′

We find that $\boldsymbol{m1}$= $\begin{bmatrix} 1 \\ -2 \end{bmatrix}$. Determine $\mathrm{T}(\boldsymbol{a2})$ and $\boldsymbol{m2}$, the second column of $\boldsymbol{M}$. Construct the matrix $\boldsymbol{M}$. Record your results below.

$\mathrm{T}(\boldsymbol{a_2}) =$ ____________ $\qquad \boldsymbol{m2} =$ ____________ $\qquad \boldsymbol{M} =$ ____________

There is a direct way to compute the matrix $\boldsymbol{M}$ in Example 1 without reference to its columns. Let $\boldsymbol{N}$ be the matrix of the transformation relative to the natural bases. Then the images of the basis vectors in $\mathcal{A}$ are the columns of $\boldsymbol{N} * \boldsymbol{A}$, so $\boldsymbol{M} = \boldsymbol{B} \setminus (\boldsymbol{N} * \boldsymbol{A})$. Determine matrix $\boldsymbol{N}$ in Example 1 and use MATLAB to compute $\boldsymbol{M}$ as indicated. Record your results below.

$\boldsymbol{N}$ =____________ $\boldsymbol{M}$ = ____________

We have described linear transformations from R^n to R^m as matrix products. The formal result is as follows. Let $\mathcal{A}$ and $\mathcal{B}$ be bases for R^n and R^m, respectively. For each vector $\boldsymbol{u}$ in R^n,

$$[T(\boldsymbol{u})]_{\mathcal{B}} = \boldsymbol{M}\,[\boldsymbol{u}]_{\mathcal{A}}$$

where $\boldsymbol{M}$ is the matrix representation of T relative to $\mathcal{A}$ and $\mathcal{B}$, $[\boldsymbol{u}]_{\mathcal{A}}$ is the coordinate vector of $\boldsymbol{u}$ relative to $\mathcal{A}$, and $[T(\boldsymbol{u})]_{\mathcal{B}}$ is the coordinate vector of T($\boldsymbol{u}$) relative to $\mathcal{B}$ (See Lab 7 page 5). In words this result says, **given the coordinates of vector $\boldsymbol{u}$ relative to basis $\mathcal{A}$ we multiply them by the matrix representation $\boldsymbol{M}$ to obtain the coordinates of the image of $\boldsymbol{u}$ relative to the $\mathcal{B}$ basis.** Basically the information in $\boldsymbol{u}$ is encoded via the $\mathcal{A}$ basis and transformed by multiplication by $\boldsymbol{M}$ into a code relative to the $\mathcal{B}$ basis. At times it may be necessary to encode a vector $\boldsymbol{u}$ given in the standard basis and decode its image into the standard basis. These actions are a preprocessing step and post processing step, respectively. It follows that the encoding step in MATLAB is

$$[\boldsymbol{u}]_{\mathcal{A}} = \mathbf{A}\backslash\mathbf{u}$$

and the decoding step is

$$\mathrm{T}(\boldsymbol{u}) = \mathbf{B}*\,[T(\boldsymbol{u})]_{\mathcal{B}}$$

A summary of results follows:

T:R^n	$\Longrightarrow$	R^m	linear trans.
$\mathcal{A}$ =$\{\boldsymbol{a1}, \boldsymbol{a2}, \ldots, \boldsymbol{an}\}$		$\mathcal{B}$ =$\{\boldsymbol{b1}, \boldsymbol{b2}, \ldots, \boldsymbol{bm}\}$	bases
A =$[\boldsymbol{a1}, \boldsymbol{a2}, \ldots, \boldsymbol{an}]$		**B** =$[\boldsymbol{b1}, \boldsymbol{b2}, \ldots, \boldsymbol{bm}]$	matrices
encoding: $[\boldsymbol{u}]_{\mathcal{A}}$ =$\mathbf{A}\backslash\mathbf{u}$		*decoding:* T($\boldsymbol{u}$) = $\mathbf{B}*\mathbf{M}*(\mathbf{A}\backslash\mathbf{u})$	

Let $\boldsymbol{N}$ be the matrix representation of T relative to the standard bases for R^n and R^m. The matrix representation of T relative to $\mathcal{A}$ and $\mathcal{B}$ is given by MATLAB statement

$$\mathbf{M} = \mathbf{B}\ \backslash(\mathbf{N}*\mathbf{A})$$

For each $\boldsymbol{u}$ in R^n the image in the standard basis is given by the MATLAB statement

$$\mathbf{Tu}{=}\mathbf{B}*\mathbf{M}*(\mathbf{A}\backslash\mathbf{u})$$

Exercises 12.2

1. Define T: $R^2 \to R^2$ by $T\left(\begin{bmatrix} x \\ y \end{bmatrix}\right) = \begin{bmatrix} x+2y \\ 2x-y \end{bmatrix}$. Let $\mathcal{A}$ be the standard basis for R^2 and let $\mathcal{B}$ be the basis $\mathcal{B} = \left\{\begin{bmatrix} -1 \\ 2 \end{bmatrix}, \begin{bmatrix} 2 \\ 0 \end{bmatrix}\right\}$. Compute the matrix representation $\boldsymbol{M}$ of T relative to each of the following pairs of bases. Record the matrix in the space provided.

a) $\mathcal{B}$ and $\mathcal{A}$. $\boldsymbol{M} =$

b) $\mathcal{A}$ and $\mathcal{B}$. $\boldsymbol{M} =$

c) $\mathcal{B}$ (and $\mathcal{B}$). $\boldsymbol{M} =$

2. Define T: $R^2 \to R^2$ by $T\left(\begin{bmatrix} x \\ y \end{bmatrix}\right) = \begin{bmatrix} x \\ -y \end{bmatrix}$. Let $\mathcal{A}$ be the standard basis for R^2 and let $\mathcal{B}$ be the basis $\mathcal{B} = \left\{\begin{bmatrix} 1 \\ -1 \end{bmatrix}, \begin{bmatrix} 1 \\ 2 \end{bmatrix}\right\}$. Compute the matrix representation $\boldsymbol{M}$ of T relative to each of the following pairs of bases. Record the matrix in the space provided.

a) $\mathcal{B}$ and $\mathcal{A}$. $\boldsymbol{M} =$

b) $\mathcal{A}$ and $\mathcal{B}$. $\boldsymbol{M} =$

c) $\mathcal{B}$ (and $\mathcal{B}$). $\boldsymbol{M} =$

3. Define T: $R^4 \to R^3$ by $T\left(\begin{bmatrix} x \\ y \\ z \\ t \end{bmatrix}\right) = \begin{bmatrix} x \\ y+z \\ z+t \end{bmatrix}$. Let $\mathcal{A} = \left\{\begin{bmatrix} 1 \\ 0 \\ 0 \\ 1 \end{bmatrix}, \begin{bmatrix} 0 \\ 0 \\ 0 \\ 1 \end{bmatrix}, \begin{bmatrix} 1 \\ 1 \\ 0 \\ 0 \end{bmatrix}, \begin{bmatrix} 0 \\ 1 \\ 1 \\ 0 \end{bmatrix}\right\}$ be a basis for R^4 and let $\mathcal{B} = \left\{\begin{bmatrix} 1 \\ 1 \\ 0 \end{bmatrix}, \begin{bmatrix} 0 \\ 1 \\ 0 \end{bmatrix}, \begin{bmatrix} 1 \\ 0 \\ 1 \end{bmatrix}\right\}$ be a basis of R^3. Compute the matrix representation $\boldsymbol{M}$ of T relative to $\mathcal{A}$ and $\mathcal{B}$. Record the matrix in the space provided.

$\boldsymbol{M}$ =

4. Let T: $R^4 \to R^3$ be the linear transformation defined by $T(\boldsymbol{u}) = \boldsymbol{N}\boldsymbol{u}$, where $\boldsymbol{N} = \begin{bmatrix} 1 & 0 & 1 & 1 \\ 0 & 1 & 2 & 1 \\ -1 & -2 & 1 & 0 \end{bmatrix}$. Let $\mathcal{A} = \left\{ \begin{bmatrix} 1 \\ 1 \\ 0 \\ 10 \end{bmatrix}, \begin{bmatrix} 0 \\ 1 \\ 0 \\ 0 \end{bmatrix}, \begin{bmatrix} 0 \\ 0 \\ 1 \\ 1 \end{bmatrix}, \begin{bmatrix} 0 \\ 1 \\ 1 \\ 0 \end{bmatrix} \right\}$ be a basis for R^4 and let $\mathcal{B} = \left\{ \begin{bmatrix} 1 \\ 0 \\ 1 \end{bmatrix}, \begin{bmatrix} 0 \\ 1 \\ 1 \end{bmatrix}, \begin{bmatrix} 0 \\ 0 \\ 1 \end{bmatrix} \right\}$ be a basis of R^3. Compute the matrix representation $\boldsymbol{M}$ of T relative to $\mathcal{A}$ and $\mathcal{B}$. Record the matrix in the space provided.

$\boldsymbol{M}$ =

5. Let $\mathcal{A}$ be the basis of R^3 defined by $\mathcal{A} = \{\boldsymbol{v1}, \boldsymbol{v2}, \boldsymbol{v3}\} = \left\{ \begin{bmatrix} 1 \\ 1 \\ 0 \end{bmatrix}, \begin{bmatrix} 0 \\ 1 \\ 1 \end{bmatrix}, \begin{bmatrix} 1 \\ 2 \\ 3 \end{bmatrix} \right\}$. Define T: $R^3 \to R^3$ by

$$T(\boldsymbol{v1}) = \begin{bmatrix} 1 \\ 2 \\ 3 \end{bmatrix}, T(\boldsymbol{v2}) = \begin{bmatrix} 4 \\ 5 \\ 6 \end{bmatrix}, T(\boldsymbol{v3}) = \begin{bmatrix} 7 \\ 8 \\ 9 \end{bmatrix}$$

Compute the matrix representation $\boldsymbol{M}$ of T relative to $\mathcal{A}$ (and $\mathcal{A}$). Record the matrix in the space provided.

$\boldsymbol{M}$ =

Section 12.3

Determining the Range in MATLAB

Let L: $\mathrm{R}^n \to \mathrm{R}^m$ be a linear transformation and let $\boldsymbol{A}$ be its matrix representation relative to the standard bases for R^n and R^m. For $\boldsymbol{x}$ in R^n the image $\mathrm{L}(\boldsymbol{x}) = \boldsymbol{Ax}$ lies in R^m. For example, suppose L: $\mathrm{R}^4 \to \mathrm{R}^3$ is a linear transformation whose matrix representation is

$$\boldsymbol{A} = \begin{bmatrix} 1 & -1 & -2 & -2 \\ 2 & -3 & -5 & -6 \\ 1 & -2 & -3 & -4 \end{bmatrix}.$$

The image of $\boldsymbol{x} = \begin{bmatrix} 1 & 2 & -1 & 0 \end{bmatrix}'$ under L is given by

$$\mathrm{L}(\boldsymbol{x}) = \boldsymbol{Ax} = \begin{bmatrix} 1 & -1 & -2 & -2 \\ 2 & -3 & -5 & -6 \\ 1 & -2 & -3 & -4 \end{bmatrix} \begin{bmatrix} 1 \\ 2 \\ -1 \\ 0 \end{bmatrix} = \begin{bmatrix} 1 \\ 1 \\ 0 \end{bmatrix}.$$

The **range** of a linear transformation L: $\mathrm{R}^n \to \mathrm{R}^m$ is the subspace of R^m consisting of all images of vectors from R^n.

It is important to observe that the vector $\boldsymbol{Ax}$ is a linear combination of the columns of $\boldsymbol{A}$. (This was the essence of Exercise 10 in Section 3.2.) For the matrix $\boldsymbol{A}$ and the vector $\boldsymbol{x}$ above,

$$\boldsymbol{Ax} = 1 \begin{bmatrix} 1 \\ 2 \\ 1 \end{bmatrix} + 2 \begin{bmatrix} -1 \\ -3 \\ -2 \end{bmatrix} - 1 \begin{bmatrix} -2 \\ -5 \\ -3 \end{bmatrix} + 0 \begin{bmatrix} -2 \\ -6 \\ -4 \end{bmatrix} = \begin{bmatrix} 1 \\ 1 \\ 0 \end{bmatrix}.$$

Since $\boldsymbol{Ax}$ is a linear combination of the columns of $\boldsymbol{A}$ it follows that

$$\text{range(L)} = \text{column space of } \boldsymbol{A}.$$

This result is critical for finding the range of a linear transformation. Thus we 'know the range of L' when we have a basis for the column space of matrix $\boldsymbol{A}$. There are two simple ways to find a basis for the column space of matrix $\boldsymbol{A}$.

- The transposes of the nonzero rows of **rref(A′)** form a basis for the column space of $\boldsymbol{A}$.

- If the columns of $\boldsymbol{A}$ containing the leading 1's of **rref(A)** are $k_1 < k_2 < \cdots < k_r$, then columns $k_1, k_2, \cdots, k_r$, form a basis for the column space of matrix $\boldsymbol{A}$.

For the matrix $\boldsymbol{A}$ above we have

$$\textbf{rref}(\mathbf{A}') = \begin{bmatrix} 1 & 0 & -1 \\ 0 & 1 & 1 \\ 0 & 0 & 0 \\ 0 & 0 & 0 \end{bmatrix}$$

and hence columns $\begin{bmatrix} 1 \\ 0 \\ -1 \end{bmatrix}$ and $\begin{bmatrix} 0 \\ 1 \\ 1 \end{bmatrix}$ are a basis for range of L. In the latter case

$$\textbf{rref}(\mathbf{A}) = \begin{bmatrix} 1 & 0 & -1 & 0 \\ 0 & 1 & 1 & 2 \\ 0 & 0 & 0 & 0 \end{bmatrix}.$$

The leading ones point to columns 1 and 2 of $\boldsymbol{A}$ as a basis for the column space of $\boldsymbol{A}$ and hence a basis for the range of L. See also MATLAB routine **lisub**, which determines a linearly independent subset of columns directly.

In summary, **rref(A)** gives sufficient information to find the range of a linear transformation.

Exercises 12.3

Find bases for the range of each linear transformation whose matrix representation is given.

Matrix Representation	Basis for Range
1. $\boldsymbol{A} = \begin{bmatrix} 1 & 2 & 5 & 5 \\ -2 & -3 & -8 & -7 \end{bmatrix}$	
2. $\boldsymbol{B} = \begin{bmatrix} -3 & 2 & -7 \\ 2 & -1 & 4 \\ 2 & -2 & 6 \end{bmatrix}$	

3. $C = \begin{bmatrix} 3 & 3 & -3 & 1 & 11 \\ -4 & -4 & 7 & -2 & -19 \\ 2 & 2 & -3 & 1 & 9 \end{bmatrix}$

Section 12.4

Determining the Kernel in MATLAB

In this section we use MATLAB to compute the kernel of a linear transformation in terms of the null space of its matrix representation. Every matrix representation in this lab is relative to standard bases.

Let L: $\mathrm{R}^n \to \mathrm{R}^m$ be a linear transformation and let $\boldsymbol{A}$ be its matrix representation relative to the standard bases for R^n and R^m. For $\boldsymbol{y}$ in R^m, each vector $\boldsymbol{x}$ in R^n such that $\mathrm{L}(\boldsymbol{x}) = \boldsymbol{y}$ is called a **pre-image** of $\boldsymbol{y}$. For example, suppose L: $\mathrm{R}^4 \to \mathrm{R}^3$ is a linear transformation whose matrix representation is

$$\boldsymbol{A} = \begin{bmatrix} 1 & -1 & -2 & -2 \\ 2 & -3 & -5 & -6 \\ 1 & -2 & -3 & -4 \end{bmatrix}.$$

Vector $\boldsymbol{x} = \begin{bmatrix} 1 & -3 & 1 & 1 \end{bmatrix}'$ is a pre-image of $\boldsymbol{y} = \begin{bmatrix} 0 & 0 & 0 \end{bmatrix}'$. This can be confirmed by verifying that $\boldsymbol{Ax} = \mathbf{0}$ or by verifying that

$$\boldsymbol{Ax} = 1 \begin{bmatrix} 1 \\ 2 \\ 1 \end{bmatrix} - 3 \begin{bmatrix} -1 \\ -3 \\ -2 \end{bmatrix} + 1 \begin{bmatrix} -2 \\ -5 \\ -3 \end{bmatrix} + 1 \begin{bmatrix} -2 \\ -6 \\ -4 \end{bmatrix} = \begin{bmatrix} 0 \\ 0 \\ 0 \end{bmatrix}.$$

The set of all pre-images of the zero vector in R^m forms a subspace of R^n, called the **kernel** of L, denoted **ker(L)**.

> The kernel of a linear transformation L: $\mathrm{R}^n \to \mathrm{R}^m$ is the subspace of R^n consisting of all vectors $\boldsymbol{x}$ such that $\mathrm{L}(\boldsymbol{x}) = \mathbf{0}$.

Since a vector $\boldsymbol{x}$ in R^n lies in the kernel of L only if

$$\mathrm{L}(\boldsymbol{x}) = \mathbf{0}$$

it follows that ker(L) is the set of all solutions of the homogeneous system

$$\boldsymbol{A}\boldsymbol{x} = \mathbf{0}.$$

The set of all solutions is often called the **null space** of the matrix representation $\boldsymbol{A}$. Thus we 'know the kernel of L' when we have a basis for the null space of the matrix representation $\boldsymbol{A}$. To find a basis for the null space of $\boldsymbol{A}$, we form the general solution of $\boldsymbol{A}\boldsymbol{x} = \mathbf{0}$ and 'separate it into a linear combination of columns using the arbitrary constants that are present.' The columns employed form a basis for the null space of $\boldsymbol{A}$. For this procedure we use the command **rref(A)**.

For example, if L: $\mathrm{R}^4 \to \mathrm{R}^3$ is a linear transformation whose matrix representation is

$$\boldsymbol{A} = \begin{bmatrix} 1 & -1 & -2 & -2 \\ 2 & -3 & -5 & -6 \\ 1 & -2 & -3 & -4 \end{bmatrix}.$$

we have

$$\mathbf{rref(A)} = \begin{bmatrix} 1 & 0 & -1 & 0 \\ 0 & 1 & 1 & 2 \\ 0 & 0 & 0 & 0 \end{bmatrix}.$$

We choose the unknowns corresponding to columns without leading 1's to be arbitrary, so set

$$x_3 = r \text{ and } x_4 = t.$$

Then $x_1 = x_3 = r$ and $x_2 = -x_3 - 2x_4 = -r - 2t$. The general solution is given by

$$\boldsymbol{x} = \begin{bmatrix} x_1 \\ x_2 \\ x_3 \\ x_4 \end{bmatrix} = \begin{bmatrix} r \\ -r-2t \\ r \\ t \end{bmatrix} = r\begin{bmatrix} 1 \\ -1 \\ 1 \\ 0 \end{bmatrix} + t\begin{bmatrix} 0 \\ -2 \\ 0 \\ 1 \end{bmatrix}$$

Hence the columns $\begin{bmatrix} 1 \\ -1 \\ 1 \\ 0 \end{bmatrix}$ and $\begin{bmatrix} 0 \\ -2 \\ 0 \\ 1 \end{bmatrix}$ form a basis for **ker(L)**. See also MATLAB routine **homsoln**, which displays the general solution of a homogeneous linear system. In addition MATLAB command **null** will produce an orthonormal basis for the null space of a matrix.

In summary, **rref(*A*)** gives sufficient information to find the kernel of a linear transformation.

Exercises 12.4

In Exercises 1 to 3, find bases for the kernel of each linear transformation whose matrix representation is given.

Matrix Representation	Basis for Kernel
1. $A = \begin{bmatrix} 1 & 2 & 5 & 5 \\ -2 & -3 & -8 & -7 \end{bmatrix}$	
2. $B = \begin{bmatrix} -3 & 2 & -7 \\ 2 & -1 & 4 \\ 2 & -2 & 6 \end{bmatrix}$	
3. $C = \begin{bmatrix} 3 & 3 & -3 & 1 & 11 \\ -4 & -4 & 7 & -2 & -19 \\ 2 & 2 & -3 & 1 & 9 \end{bmatrix}$	

4. Let $\boldsymbol{A} = \begin{bmatrix} 1 & 2 & 4 & -2 \\ 2 & 1 & 2 & 0 \\ 0 & 3 & 6 & -4 \end{bmatrix}$ and $\mathrm{L}(\boldsymbol{x}) = \boldsymbol{A}\boldsymbol{x}$.

a) Find a basis for the kernel of L or equivalently the null space of $\boldsymbol{A}$. List the basis below.

b) Find a basis for the row space of $\boldsymbol{A}$. List the basis below.

c) Compute the dot product of each of the basis vectors of the null space of $\boldsymbol{A}$ with each basis vector of the row space of $\boldsymbol{A}$. Write a description of the results below.

d) If $\boldsymbol{w}$ is in the null space of $\boldsymbol{A}$ and $\boldsymbol{v}$ is in the row space of $\boldsymbol{A}$, then what is **dot(w,v)**? Explain your answer.

e) Fill in the blank: Every vector in the row space of $\boldsymbol{A}$ is ____________ to every vector in the null space of $\boldsymbol{A}$.

f) Are there any vectors in the row space of $\boldsymbol{A}$ (considered as columns) that are also in the null space of $\boldsymbol{A}$? Explain your answer.

5. Let $\boldsymbol{A} = \begin{bmatrix} 5 & 5 & 3 \\ -7 & -7 & -3 \\ 5 & 5 & 3 \end{bmatrix}$ and define $\mathrm{L}(\boldsymbol{x}) = (\boldsymbol{A} - t\boldsymbol{I})\boldsymbol{x}$.

a) Experiment to find a positive integer t so that the kernel of L is not the zero vector alone. Then find a basis for the kernel using the value of t you found.

b) Let $\boldsymbol{y} = -5*$ your basis vector from part a and compute $\boldsymbol{z} = \boldsymbol{A} * \boldsymbol{y}$. Express $\boldsymbol{z}$ in terms of t and your basis vector. Geometrically what is the relationship between $\boldsymbol{y}$ and $\boldsymbol{z}$?

c) Repeat part a), except find a negative integer t.

d) Let $\boldsymbol{q}= 8*$ your basis vector from part c) and compute $\boldsymbol{s} = \boldsymbol{A} * \boldsymbol{q}$. Express $\boldsymbol{s}$ in terms of t and your basis vector. Geometrically what is the relationship between $\boldsymbol{q}$ and $\boldsymbol{s}$?

e) For $t = 0$ the kernel of L is not the zero vector alone. Find a nonzero vector $\boldsymbol{r}$ in the kernel of $\mathrm{L}(\boldsymbol{x})$ in this case.

f) Let $\boldsymbol{P}$ be the 3×3 matrix whose first column is the basis vector from part a), second column is the basis vector from part c) and third column is the vector from part e). Compute $\boldsymbol{Q} = \boldsymbol{P}^{-1}\boldsymbol{A}\boldsymbol{P}$. Describe the relationship between $\boldsymbol{Q}$ and other parts of this problem.

g) The matrix $\boldsymbol{Q}$ in part f) is a(n) ________________ matrix.

<< **NOTES; COMMENTS; IDEAS** >>

LAB 13

The Eigenproblem

Topics: routines **matvec**, **evecsrch** and **mapcirc**; eigenvalues, eigenvectors, characteristic polynomial, roots of the characteristic polynomial; applications.

Introduction

This lab contains both a geometric and an algebraic development of eigen concepts. The geometric development in Section 13.1 uses the function $f(\boldsymbol{x}) = \boldsymbol{A}\boldsymbol{x}$ where $\boldsymbol{A}$ is a 2×2 matrix and $\boldsymbol{x}$ is a vector in R^2. We compare the input $\boldsymbol{x}$ and output $\boldsymbol{A}\boldsymbol{x}$ graphically using routine **matvec**. This routine employs MATLAB 's graphical user interface to actively engage the student in experiments which provide a foundation for basic eigen concepts. We follow with routine **evecsrch**, which automates the manual search of **matvec**. Section 13.1 requires only matrix algebra, the notion of length of a vector, and linear independence. It can be used before determinants have been developed. Several exercises explore $f(\boldsymbol{x}) = \boldsymbol{A}\boldsymbol{x}$ as a linear transformation using the image of the unit circle within another routine **mapcirc**.

Section 13.2 develops the algebraic solution of $f(\boldsymbol{x}) = \boldsymbol{A}\boldsymbol{x} = \lambda\boldsymbol{x}$ using MATLAB commands for determinants, the characteristic polynomial and its roots, and reduced row echelon form, then concludes with an introduction to MATLAB 's command for eigen computation. This section does not depend on Section 13.1 so it can be used to emphasize the algebraic aspects of the eigen problem. With Section 13.1 this section provides the complement of the geometric exploration.

Section 13.3 provides a set of experiments to explore properties of eigenvalues and eigenvectors, diagonalizable matrices, and applications involving matrix powers, Markov population models, and graph theory applied to a geographical problem involving trade routes.

Section 13.1

Discussion of the General Concept

From a general point of view an eigenvalue is an 'item' in a situation (in mathematics we say problem) that must take on specific values in order for the situation to have a desired outcome. The 'item' (sometimes referred to as a design parameter) is often a constant that can be changed or tuned in an effort to produce a desired outcome (or solution) for the situation. From the description of the situation we must develop an appropriate strategy in order to determine a setting (in mathematics we say value) for the eigenvalue so we can achieve the desired outcome. Hence the determination of an eigenvalue is required as a step of a process to achieve a desired outcome (solution). It follows that we must solve for an eigenvalue to ensure that the desired solution exists. Unfortunately, more than one setting (value) for the eigenvalue could lead to a solution, but not all such settings necessarily produce the outcome we desire.

Consider designing a clock mechanism by using a mass M attached to a spring which is fixed to a support. We desire to have the mass-spring system vibrate so that at 1 second intervals the mass is back at its original position to trigger the movement of a second hand. At time t let the distance of the mass from its original position be denoted by $x(t)$. Then a simple mathematical model for this situation is given by the differential equation

$$\frac{d^2x(t)}{dt^2} + \lambda x(t) = 0$$

where we have constraints $x(0) = 0, x(1) = 0$. (The constraints are called boundary values.) It is known that the eigenvalue λ depends on physical characteristics of the spring and the size of the mass. It happens that there are infinitely many values of λ that will produce a solution for this problem, but not all such values of λ give a solution that we want to use in the design of our clock. For instance, $\lambda = 0$ gives the solution $x(t) = 0$; that is, the mass never moves. Hence the second hand remains fixed. Certainly this choice for the eigenvalue λ gives a valid mathematical solution to the problem, but the corresponding solution is unacceptable if we hope to sell clocks. **An acceptable solution is called an eigenvector (in this case an eigenfunction) associated with the eigenvalue used to obtain that solution.**

Another situation where we vary an 'item' to achieve a result involves tuning in a particular radio station. The tuning knob of a radio varies the capacitance in the tuning circuit. In this way the resonant frequency is changed until it agrees with the frequency of the station we desire. In a broad sense an eigenvalue was selected by the tuning knob to produce a desired solution.

The basic equation that arises to compute eigenvalues λ and corresponding eigenvectors $\boldsymbol{x}$ is

$$f(\boldsymbol{x}) = \lambda \boldsymbol{x}$$

The situation determines the details of the function f. Intuitively this equation says that we seek an *input* $\boldsymbol{x}$ such that *output* $f(\boldsymbol{x})$ is a scalar multiple of $\boldsymbol{x}$. A simple geometric interpretation is that $\boldsymbol{x}$ is an eigenvector of f provided that output $f(\boldsymbol{x})$ is parallel to input $\boldsymbol{x}$. The associated eigenvalue λ is viewed as a magnification factor which affects the direction and length of the output. An easy model of this situation is to consider $f(\boldsymbol{x}) = \boldsymbol{A}\boldsymbol{x}$ in which $\boldsymbol{A}$ is a 2×2 matrix and $\boldsymbol{x}$ is a vector in R^2. In this case the eigen equation is

$$\boldsymbol{A}\boldsymbol{x} = \lambda \boldsymbol{x}$$

So we seek vectors $\boldsymbol{x}$ so that vector $\boldsymbol{y} = \boldsymbol{A}\boldsymbol{x}$ is parallel to $\boldsymbol{x}$. To illustrate this graphically we use MATLAB routine **matvec**. For a user-chosen 2×2 matrix $\boldsymbol{A}$ use the mouse to choose an input vector $\boldsymbol{x}$ from the unit circle. This input vector is displayed graphically, then the output vector $\boldsymbol{y} = \boldsymbol{A}\boldsymbol{x}$ is computed, scaled to have length 1, and displayed graphically. The routine allows multiple selections of inputs $\boldsymbol{x}$ so you can 'home-in' on an eigenvector.

Example 1. In MATLAB type **matvec**. From the menu displayed choose the option for the built-in demo. Your screen should look like Figure 1.

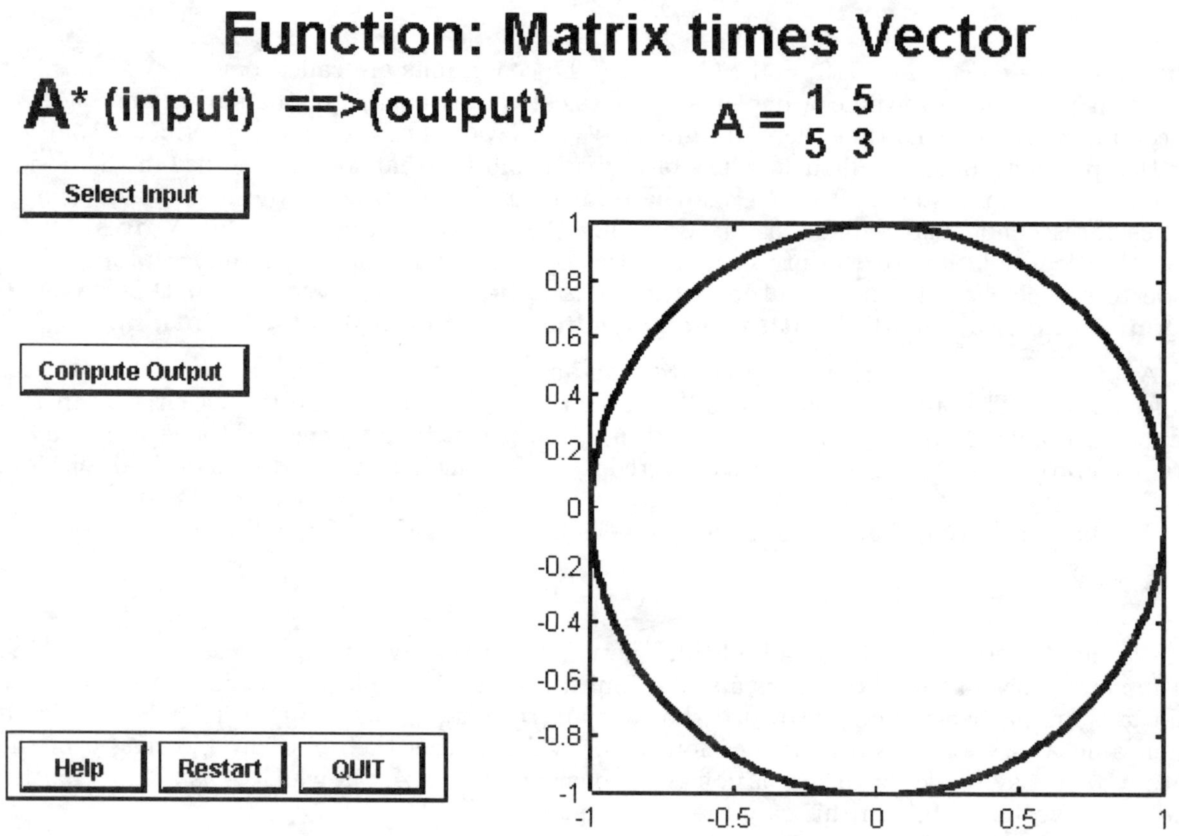

Figure 1.

Click on the button **Select Input**. A message will appear directing you to click on (the circumference of the) circle at the right to select input $\boldsymbol{x}$. After you make your selection the coordinates of $\boldsymbol{x}$ are displayed and the vector drawn from the center of the unit circle.

Next click on the button **Compute Output**. The coordinates of $\boldsymbol{y} = \boldsymbol{A}\boldsymbol{x}$ are shown together with the coordinates of the scaled output vector and its graphical representation.

The **More** button, which appears after the execution of the 'Compute Output', encourages further experimentation. Click on this button. Now click once again on 'Select Input'. A small circle remains on the circumference of the unit circle to indicate where previous inputs were chosen.

Use the mouse to choose inputs until you 'home-in' on an eigenvector of matrix $\boldsymbol{A}$. (Hint: The origin is placed at the center of the circle. Choose inputs in the first quadrant.)

Once you have a close approximation to an eigenvector record the coordinates below.

$$\text{input} = \begin{bmatrix} \_\_\_\_ \\ \_\_\_\_ \end{bmatrix} \quad \text{output} = \begin{bmatrix} \_\_\_\_ \\ \_\_\_\_ \end{bmatrix} \quad \text{scaled output} = \begin{bmatrix} \_\_\_\_ \\ \_\_\_\_ \end{bmatrix}$$

(By close approximation to an eigenvector we mean geometrically that the input and output are nearly parallel and algebraically that the first few decimal places of the coordinates of the input and scaled output are the same.)

Search for a second eigenvector by choosing input vectors in the second quadrant. Once you have a close approximation to an eigenvector record the coordinates below.

$$\text{input} = \begin{bmatrix} \_\_\_\_ \\ \_\_\_\_ \end{bmatrix} \quad \text{output} = \begin{bmatrix} \_\_\_\_ \\ \_\_\_\_ \end{bmatrix} \quad \text{scaled output} = \begin{bmatrix} \_\_\_\_ \\ \_\_\_\_ \end{bmatrix}$$

To exit **matvec** click on the QUIT button.

In Example 1 matrix $\boldsymbol{A} = \begin{bmatrix} 1 & 5 \\ 5 & 3 \end{bmatrix}$. From the graphical displays you should have observed that for the eigenvector $\boldsymbol{x}_1$ in the first quadrant, the output $\boldsymbol{A}\boldsymbol{x}_1$ was also in the same quadrant. This implies that the corresponding eigenvalue is positive. The numerical value of corresponding eigenvalue λ_1 is obtained from the basic equation $\boldsymbol{A}\boldsymbol{x}_1 = \boldsymbol{\lambda}_1\boldsymbol{x}_1$ by taking the norm of both sides and solving for λ_1;

$$| \lambda_1 | = \frac{norm(\boldsymbol{A} * \boldsymbol{x}_1)}{norm(\boldsymbol{x}_1)}$$

But here $\boldsymbol{x}_1$ is a unit vector so the denominator is 1 and we have seen from geometric considerations that this eigenvalue is positive so

$$\lambda_1 = norm(\boldsymbol{A} * \boldsymbol{x}_1)$$

The eigenvector $\boldsymbol{x}_1$ is approximately $\begin{bmatrix} 0.6340 \\ 0.7733 \end{bmatrix}$. Estimate the corresponding eigenvalue λ_1.

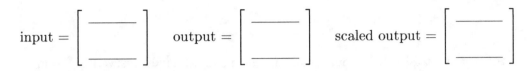

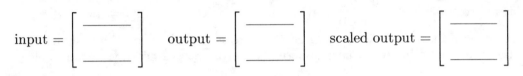

Record your work below.

$\lambda_1 =$______

Let $\boldsymbol{x_2}$ denote the eigenvector in the second quadrant. You should have observed in Example 1 that output $\boldsymbol{Ax_2}$ was in the fourth quadrant. In the space below give an argument to verify that the corresponding eigenvalue λ_2 must be negative.

Eigenvector $\boldsymbol{x_2}$ is approximately $\begin{bmatrix} 0.7733 \\ -0.6340 \end{bmatrix}$. Once again, the command $\lambda_2 = norm(\boldsymbol{A} * \boldsymbol{x_2})$ yields the absolute value of the second eigenvalue. But geometric considerations have shown that this value must be negative. Estimate the corresponding eigenvalue λ_2. Record your work below.

$\lambda_2 =$______

The routine **matvec** used in Example 1 lets you experiment to determine eigenvectors of a 2×2 matrix. By clicking on the unit circle you really chose a direction for the input vector $\boldsymbol{x}$. The calculation of the output vector $\boldsymbol{y} = \boldsymbol{Ax}$ determines another direction. We say we have an *eigen direction* of $\boldsymbol{A}$ provided the input and output directions are parallel. That is, input and output are in exactly the same direction or in exactly opposite directions. The use of the unit circle is a convenience, really any circle centered at the origin could be used. To emphasize this, in the space below give an argument that verifies that if $\boldsymbol{x}$ is an eigenvector of $\boldsymbol{A}$, that is $\boldsymbol{Ax} = \lambda\boldsymbol{x}$, then $k{*}\boldsymbol{x}$ is an eigenvector for any scalar $k \neq 0$.

(The preceding shows that the set of eigenvectors of $\boldsymbol{A}$ corresponding to an eigenvalue λ is closed under scalar multiplication.) Thus it is the *'direction'* that is important for an eigenvector, not its length. That is why we could scale the output in **matvec** and display it on the unit circle.

For a 2×2 matrix $\boldsymbol{A}$ the search for eigenvectors can be illustrated by selecting vectors that encompass directions around the unit circle and checking to see if the *input* and *output* = $\boldsymbol{A}{*}$*input* are parallel. Routine **evecsrch** automates this process. The search starts by selecting an input at a randomly chosen point on the unit circle and graphing the corresponding radius.

Next the output is computed, scaled to the unit circle, and graphed. If the input and output are parallel the images are retained, otherwise both are erased. When an eigenvector is detected its components are displayed. The routine stops when it completes the search of the 'entire' unit circle.

In MATLAB type **evecsrch** and follow the directions. Use **evecsrch** to determine the eigenvectors of $\boldsymbol{A} = \begin{bmatrix} 0 & 6 \\ 1 & -1 \end{bmatrix}$. Record the eigenvectors below.

By observation, explain why these eigenvectors are linearly independent.

Determine the corresponding eigenvalues. Show your work below.

In the space below explain why the radii drawn form an **'X'** on the unit circle in the output from **evecsrch**.

Use **evecsrch** to investigate the eigenvectors of $\boldsymbol{B} = \begin{bmatrix} 6 & 7 \\ 0 & 6 \end{bmatrix}$ and determine the corresponding eigenvalues. In the space below summarize your observations and record your calculations.

Next compare the results of **evecsrch** for the preceding matrices $\boldsymbol{A}$ and $\boldsymbol{B}$ in terms of eigenvector information. What is different? Put your discussion below.

Finally, use **evecsrch** to search for the eigenvectors of $\boldsymbol{A} = \begin{bmatrix} 1 & 5 \\ 5 & 3 \end{bmatrix}$, which is the matrix from Example 1. Verify that the eigenvectors are the ones that were computed there.

Exercises 13.1

1. Let $\boldsymbol{A} = \begin{bmatrix} 6 & 1 \\ -2 & 0 \end{bmatrix}$.

a) Use routine **evecsrch** to approximate eigenvectors $\boldsymbol{x}_1$ and $\boldsymbol{x}_2$ of $\boldsymbol{A}$.

$$\boldsymbol{x}_1 = \begin{bmatrix} \_\_\_\_\_ \\ \_\_\_\_\_ \end{bmatrix} \qquad \boldsymbol{x}_2 = \begin{bmatrix} \_\_\_\_\_ \\ \_\_\_\_\_ \end{bmatrix}$$

b) Use routine **matvec** to determine the output when the input is selected close to $\boldsymbol{x}_1$.

Is the output in the same direction as $\boldsymbol{x}_1$, or in the opposite direction? ____________

c) From part b it follows that the eigenvalue λ_1 corresponding to $\boldsymbol{x}_1$ is (circle one):

Positive Negative

d) Combine parts a and c with command **norm(A*x$_1$)** to approximate λ_1.

$\lambda_1 \approx$ ____________

e) Use routine **matvec** to determine the output when the input is selected close to $\boldsymbol{x}_2$.

Is the output in the same direction as $\boldsymbol{x}_2$, or in the opposite direction? ____________

f) Combine parts a and e with command **norm(A*x$_2$)** to approximate λ_2.

$\lambda_2 \approx$ ____________

g) Use MATLAB to determine whether $\boldsymbol{A} * \boldsymbol{x}_1 = \lambda_1 \boldsymbol{x}_1$ and $\boldsymbol{A} * \boldsymbol{x}_2 = \lambda_2 \boldsymbol{x}_2$. Explain any discrepancies in the space below.

2. Let $\boldsymbol{A} = \begin{bmatrix} 1 & 9 \\ 1 & 1 \end{bmatrix}$.

a) Use routine **evecsrch** to approximate eigenvectors $\boldsymbol{x}_1$ and $\boldsymbol{x}_2$ of $\boldsymbol{A}$.

$$\boldsymbol{x}_1 = \begin{bmatrix} \_\_\_\_\_ \\ \_\_\_\_\_ \end{bmatrix} \qquad \boldsymbol{x}_2 = \begin{bmatrix} \_\_\_\_\_ \\ \_\_\_\_\_ \end{bmatrix}$$

b) Use routine **matvec** to determine the output when the input is selected close to $\boldsymbol{x}_1$ that lies in the third quadrant.

Is the output in the same direction as $\boldsymbol{x}_1$ or in the opposite direction? ____________

c) Combine parts a and b with command **norm(A*x$_1$)** to approximate λ_1.

$\lambda_1 \approx$ ____________

d) Use routine **matvec** to determine the output when the input is selected close to $\boldsymbol{x}_2$.

Is the output in the same direction as $\boldsymbol{x}_2$, or in the opposite direction? ____________

e) Combine parts a and d with command **norm(A*x$_2$)** to approximate λ_2.

$\lambda_2 \approx$ ____________

f) Use MATLAB to determine whether $\boldsymbol{A} * \boldsymbol{x}_1 = \lambda_1 \boldsymbol{x}_1$ and $\boldsymbol{A} * \boldsymbol{x}_2 = \lambda_2 \boldsymbol{x}_2$. Explain any discrepancies in the space below.

3. Use routine **matvec** to approximate an eigenvector of $\boldsymbol{A} = \begin{bmatrix} 2 & 3 \\ 0 & 0 \end{bmatrix}$ that is in the first quadrant. Record your approximate eigenvector $\begin{bmatrix} \_\_\_\_ \\ \_\_\_\_ \end{bmatrix}$.

4. Use the matrix $\boldsymbol{A}$ from Exercise 3.

a) Algebraically find an eigenvector $\boldsymbol{x} = \begin{bmatrix} x_1 \\ x_2 \end{bmatrix}$ by solving the matrix equation $\boldsymbol{Ax} = 2\boldsymbol{x}$ for x_1 and x_2. Show your work in the space below.

b) How many solutions were there to the matrix equation $\boldsymbol{Ax} = 2\boldsymbol{x}$ in part a?

____

c) Take your solution from part a and scale it to have length 1. Record that vector below.

$$\begin{bmatrix} \_\_\_\_ \\ \_\_\_\_ \end{bmatrix}$$

d) Find another eigenvector of length 1 which is parallel to your solution in part c. Record that vector below.

$$\begin{bmatrix} \_\_\_\_ \\ \_\_\_\_ \end{bmatrix}$$

e) What solution from part a cannot be scaled to have length 1? Explain.

5. Let $\boldsymbol{A} = \begin{bmatrix} 5 & -1 \\ -1 & 5 \end{bmatrix}$.

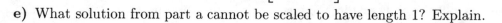

a) Use **matvec** to approximate an eigenvector of $\boldsymbol{A}$ that is in the first quadrant. Record that vector below. (Make selections until at least the first two decimal places in the input and scaled output vectors agree.)

$$\begin{bmatrix} \_\_\_\_\_\_ \\ \_\_\_\_\_\_ \end{bmatrix}$$

b) For this matrix there is a second eigenvector that is orthogonal to the one you found in part a. Determine this vector using **matvec**. Record that vector below. (Make selections until at least the first two decimal places in the input and scaled output vectors agree.)

$$\begin{bmatrix} \_\_\_\_\_\_ \\ \_\_\_\_\_\_ \end{bmatrix}$$

Verify that the vector in part a is orthogonal to the vector you found here. (Actually it may only be close to orthogonal with the vector in part a since we are using agreement in only the first two decimal places to indicate the input and scaled output vectors are parallel.)

6. Let $\boldsymbol{A}$ be a 2×2 diagonal matrix. Perform a set of experiments using routine **evecsrch** to determine the eigenvectors and corresponding eigenvalues of $\boldsymbol{A}$. Below each of the following matrices record your findings.

$$\begin{bmatrix} 5 & 0 \\ 0 & 3 \end{bmatrix} \qquad \begin{bmatrix} -8 & 0 \\ 0 & 1 \end{bmatrix} \qquad \begin{bmatrix} 4 & 0 \\ 0 & 4 \end{bmatrix}$$

Perform additional experiments to formulate a conjecture that describes the eigenvalues and corresponding eigenvectors of a 2×2 diagonal matrix.

Conjecture: ____

____

7. Let $\boldsymbol{A}$ be a 2×2 upper triangular matrix . Perform a set of experiments using routine **evecsrch** to determine the eigenvectors and corresponding eigenvalues of $\boldsymbol{A}$. Below each of the following matrices record your findings.

$$\begin{bmatrix} -5 & 4 \\ 0 & 2 \end{bmatrix} \qquad \begin{bmatrix} 8 & 1 \\ 0 & 1 \end{bmatrix} \qquad \begin{bmatrix} -4 & 3 \\ 0 & 4 \end{bmatrix}$$

Perform additional experiments to formulate a conjecture that describes the eigenvalues and corresponding eigenvectors of a 2×2 upper triangular matrix.

Conjecture: ______________________________

____

8. Let $\boldsymbol{A}$ be a 2×2 lower triangular matrix . Perform a set of experiments using routine **evecsrch** to determine the eigenvectors and corresponding eigenvalues of $\boldsymbol{A}$. Below each of the following matrices record your findings.

$$\begin{bmatrix} -7 & 0 \\ 6 & 2 \end{bmatrix} \qquad \begin{bmatrix} -9 & 0 \\ 3 & 1 \end{bmatrix} \qquad \begin{bmatrix} -2 & 0 \\ 2 & 4 \end{bmatrix}$$

Perform additional experiments to formulate a conjecture that describes the eigenvalues and corresponding eigenvectors of a 2×2 lower triangular matrix.

Conjecture: ______________________________

____

9. Let $\boldsymbol{A}$ be a 2×2 matrix with a zero row. Perform a set of experiments using routine **evecsrch** to determine the eigenvectors and corresponding eigenvalues of $\boldsymbol{A}$. Below each of the following matrices record your findings.

$$\begin{bmatrix} -5 & 4 \\ 0 & 0 \end{bmatrix} \qquad \begin{bmatrix} 0 & 0 \\ 6 & 1 \end{bmatrix} \qquad \begin{bmatrix} -4 & 3 \\ 0 & 0 \end{bmatrix}$$

Perform additional experiments to formulate a conjecture that describes the eigenvalues and corresponding eigenvectors of a 2×2 matrix with a zero row.

Conjecture: ____

____

10. Let $\boldsymbol{A}$ be a 2×2 matrix with a zero column. Perform a set of experiments using routine **evecsrch** to determine the eigenvectors and corresponding eigenvalues of $\boldsymbol{A}$. Below each of the following matrices record your findings.

$$\begin{bmatrix} -5 & 0 \\ 6 & 0 \end{bmatrix} \qquad \begin{bmatrix} 0 & 9 \\ 0 & 1 \end{bmatrix} \qquad \begin{bmatrix} -4 & 0 \\ 0 & 0 \end{bmatrix}$$

Perform additional experiments to formulate a conjecture that describes the eigenvalues and corresponding eigenvectors of a 2×2 matrix with a zero column.

Conjecture: ____

____

11. Let $\boldsymbol{A}$ be a 2×2 symmetric matrix. Perform a set of experiments using routine **evecsrch** to determine an algebraic relationship between the eigenvectors of $\boldsymbol{A}$. Record your findings below each of the following matrices.

$$\begin{bmatrix} -5 & 9 \\ 9 & 0 \end{bmatrix} \qquad \begin{bmatrix} 0 & -2 \\ -2 & 1 \end{bmatrix} \qquad \begin{bmatrix} -4 & 6 \\ 6 & 0 \end{bmatrix}$$

Perform additional experiments to formulate a conjecture that describes this algebraic relationship between the eigenvectors of a 2×2 symmetric matrix.

Conjecture: ____

____

Formulate a geometric analog of the algebraic relationship.

Geometric formulation: ______________________________

____

In **matvec** we selected input vectors $\boldsymbol{x}$ from the unit circle and displayed their image $\boldsymbol{y} = \boldsymbol{A} * \boldsymbol{x}$ scaled to the unit circle. In **evecsrch** we searched for inputs around the unit circle so that the images would be parallel to the input vector. In Exercises 12 - 19 we look at the entire set of images of the unit circle. We investigate the image of the unit circle geometrically and provide experiments to investigate properties of this image for certain types of 2×2 matrices.

Before performing the investigations below execute routine **mapcirc** which is our primary investigation tool. In MATLAB type **mapcirc**. From the first menu select the built-in demonstration. From the second menu select *not* to see eigenvector information. In the display generated by **mapcirc** the left graph shows the unit circle and the right graph its image under the mapping (or transformation) by matrix $\boldsymbol{A}$. As vectors $\boldsymbol{x}$ are selected from the unit circle on the left their image $\boldsymbol{A} * \boldsymbol{x}$ is computed and displayed on the right graph. For more information on **mapcirc** use **help**.

12. Use **mapcirc** to obtain the image of the unit circle for each of the following matrices. Below each matrix record a sketch of the image and give a geometric description on the line provided.

$$\begin{bmatrix} 3 & 0 \\ 0 & 3 \end{bmatrix} \qquad \begin{bmatrix} 1 & 0 \\ 0 & 4 \end{bmatrix} \qquad \begin{bmatrix} 3 & 0 \\ 0 & -1 \end{bmatrix}$$

______________ ______________ ______________

$$\begin{bmatrix} -1/2 & 0 \\ 0 & -1/2 \end{bmatrix} \qquad \begin{bmatrix} 1 & 0 \\ 0 & 6 \end{bmatrix} \qquad \begin{bmatrix} 4.5 & 0 \\ 0 & 1 \end{bmatrix}$$

______________ ______________ ______________

Based on the previous experiments form a conjecture about the shape of the image of the unit circle when $\boldsymbol{A}$ has the following forms.

For $\boldsymbol{A} = \begin{bmatrix} k & 0 \\ 0 & k \end{bmatrix}$ Conjecture: The image of the unit circle is

____

For $\boldsymbol{A} = \begin{bmatrix} 1 & 0 \\ 0 & n \end{bmatrix}$ Conjecture: The image of the unit circle is

____

For $\boldsymbol{A} = \begin{bmatrix} m & 0 \\ 0 & 1 \end{bmatrix}$ Conjecture: The image of the unit circle is

____

13. Use **mapcirc** to obtain the image of the unit circle for each of the following matrices. Below each matrix record a sketch of the image and give a geometric description on the line provided.

$\begin{bmatrix} 1 & 2 \\ 0 & 1 \end{bmatrix}$ $\begin{bmatrix} 1 & 4 \\ 0 & 1 \end{bmatrix}$ $\begin{bmatrix} 1 & 6 \\ 0 & 1 \end{bmatrix}$

__________ __________ __________

$\begin{bmatrix} 1 & 0 \\ 2 & 1 \end{bmatrix}$ $\begin{bmatrix} 1 & 0 \\ 4 & 1 \end{bmatrix}$ $\begin{bmatrix} 1 & 0 \\ 6 & 1 \end{bmatrix}$

__________ __________ __________

Based on the previous experiments form a conjecture about the shape of the image of the unit circle when $\boldsymbol{A}$ has the following forms.

For $\boldsymbol{A} = \begin{bmatrix} 1 & k \\ 0 & 1 \end{bmatrix}$ with $k > 0$ and large Conjecture: The image of the unit circle is

____

For $\boldsymbol{A} = \begin{bmatrix} 1 & 0 \\ k & 1 \end{bmatrix}$ with $k > 0$ and large Conjecture: The image of the unit circle is

____

14. This exercise continues the type of investigation begun in the preceding exercise.

a) Design a set of experiments to investigate the behavior of the images of the unit circle when $\boldsymbol{A}$ has the form $\begin{bmatrix} 1 & k \\ 0 & 1 \end{bmatrix}$ with $k < 0$ and $|\,k\,|$ large. Summarize the behavior in a short paragraph below.

How does the behavior here differ from the case in the preceding exercise where $k > 0$? (Be specific.)

____

b) Design a set of experiments to investigate the behavior of the images of the unit circle when $\boldsymbol{A}$ has the form $\begin{bmatrix} 1 & 0 \\ k & 1 \end{bmatrix}$ with $k < 0$ and $|\,k\,|$ large. Summarize the behavior in a short paragraph below.

How does the behavior here differ from the case in the preceding exercise where $k > 0$? (Be specific.)

____

15. Use **mapcirc** to obtain the image of the unit circle for each of the following matrices. Below each matrix record a sketch of the image and give a geometric description on the line provided.

$$\begin{bmatrix} 5 & 1 \\ 5 & 1 \end{bmatrix} \qquad \begin{bmatrix} -1 & 2 \\ 1 & -2 \end{bmatrix} \qquad \begin{bmatrix} 3 & 4 \\ 0 & 0 \end{bmatrix} \qquad \begin{bmatrix} -2 & 0 \\ 1 & 0 \end{bmatrix}$$

__________ __________ __________ __________

From the geometric description above, what is the dimension of the null space of each of these matrices?

____

Each of these matrices is (circle one) singular nonsingular

Conjecture: The image of the unit circle by a _______________ matrix is a _______________.

If $\boldsymbol{A} = \begin{bmatrix} 0 & 0 \\ 0 & 0 \end{bmatrix}$ what is the image of the unit circle?

____

Is your conjecture immediately above still true? Revise it if necessary.

16. Design a set of experiments to develop a conjecture concerning the image of the unit circle if $\boldsymbol{A}$ is a 2×2 nonsingular matrix.

Conjecture:____

17. Based on the previous exercises, complete the following sentence:
The image of a unit circle under a 2×2 matrix is a __________, a __________, or a __________.

18. For each of the following 2×2 symmetric matrices use **mapcirc** to graphically determine the eigenvectors from their images. On the lines below each matrix first describe the angle between the eigenvectors and on the second line the angle between their images.

$$\begin{bmatrix} 1 & 2 \\ 2 & -2 \end{bmatrix} \qquad \begin{bmatrix} 3 & 5 \\ 5 & 0 \end{bmatrix} \qquad \begin{bmatrix} -2 & -1 \\ -1 & 4 \end{bmatrix}$$

____________ ____________ ____________

____________ ____________ ____________

Complete the following conjectures:

a) The eigenvectors of a symmetric matrix are ____________________.

b) The images of the eigenvectors of a symmetric matrix are ____________________.

c) The images of the eigenvectors of a symmetric matrix form the ____________________

and ____________________ of the elliptical image.

19. For each of the following symmetric matrices use **mapcirc** to determine the images of the eigenvectors. When **mapcirc** is over return to the command screen and type the following commands. (Choose the option to display the eigenvectors.)

pts = ginput(2); image=pts'

You will be returned to the graphics screen generated by **mapcirc**. The mouse pointer symbol will be a plus sign indicating that you can collect information about points. *Carefully* position the plus sign at the end points of the images of the eigenvectors of $\boldsymbol{A}$ which appear in a contrasting color. Click the mouse to record the coordinates of those points. After the second click press ENTER to return to the command screen. The coordinates of the end points of the vectors you clicked on will be the columns displayed in matrix **image**. (If you feel you did not position the mouse correctly for the preceding measurements, just repeat the commands above.)

Compute the length of each column of **image** using the following commands.

L1 = norm(image(:,1)),L2=norm(image(:,2))

Record these values on the lines provided and then record the closest integer to the value.

a) $A = \begin{bmatrix} 2.6 & -0.8 \\ -0.8 & 1.4 \end{bmatrix}$ L1 = ____________, L2 = ____________

b) $A = \begin{bmatrix} -3.6 & 2.8 \\ 2.8 & -0.6 \end{bmatrix}$ L1 = ____________, L2 = ____________

c) $A = \begin{bmatrix} 3.12 & 3.84 \\ -3.84 & 0.88 \end{bmatrix}$ L1 = ____________, L2 = ____________

Form a conjecture that describes the values of L1 and L2. (Hint: review Example 1 and the discussion following it.)

Conjecture: ____________________________________

What do the lengths L1 and L2 represent geometrically in terms of the image of the unit circle which is an ellipse?

Conjecture: ____________________________________

Section 13.2

The Matrix Eigenproblem

We study a surprisingly simple problem involving eigenvalues and eigenvectors that extends the 2×2 matrices considered in Section 1. Given an $n \times n$ matrix $\boldsymbol{A}$, determine how to select vectors $\boldsymbol{x}$ in R^n so that $\boldsymbol{Ax}$ is parallel to $\boldsymbol{x}$. In the context of linear transformations, let L: $\mathrm{R}^n \to \mathrm{R}^n$ be defined by

$$\mathrm{L}(\boldsymbol{x}) = \boldsymbol{Ax}.$$

We seek input vectors $\boldsymbol{x}$ that are parallel to the output vector $\boldsymbol{Ax}$. To determine a mathematical model for this problem, recall that two vectors are parallel provided they are scalar multiples of one another. Thus our objective is to find $\boldsymbol{x}$ in R^n so that

$$\mathrm{L}(\boldsymbol{x}) = \boldsymbol{Ax} = \lambda\boldsymbol{x}$$

where λ is some scalar. In the matrix equation $\boldsymbol{Ax} = \lambda\boldsymbol{x}$, both the vector $\boldsymbol{x}$ and the scalar λ are unknown. By observation we see that one solution is $\boldsymbol{x} = \boldsymbol{0}$ and then λ could be any value. This solution is uninteresting since $\boldsymbol{A0} = \boldsymbol{0}$ implies only that the zero vector is parallel to itself. Thus we exclude $\boldsymbol{x} = \boldsymbol{0}$ from acceptable solutions. We state our problem as follows.

<u>The Eigenproblem for an $n \times n$ matrix $\boldsymbol{A}$</u>

Determine a nonzero vector $\boldsymbol{x}$ in R^n and scalar λ so that $\boldsymbol{Ax} = \lambda\boldsymbol{x}$. We say λ is an **eigenvalue** of matrix $\boldsymbol{A}$ and $\boldsymbol{x}$ is an associated **eigenvector**.

A basic strategy to compute eigenvalues and eigenvectors of a matrix $\boldsymbol{A}$ begins with the matrix equation $\boldsymbol{Ax} = \lambda\boldsymbol{x}$ and uses concepts we studied previously. We have the following set of equivalent expressions:

$$\boldsymbol{Ax} = \lambda\boldsymbol{x} \Longleftrightarrow \boldsymbol{Ax} = \lambda\boldsymbol{I_n x} \Longleftrightarrow \boldsymbol{Ax}-\lambda\boldsymbol{I_n x} = \boldsymbol{0} \Longleftrightarrow (\boldsymbol{A}-\lambda\boldsymbol{I_n})\boldsymbol{x} = \boldsymbol{0}$$

Thus our eigenproblem has been recast as a homogeneous system of equations $(\boldsymbol{A}-\lambda\boldsymbol{I})\boldsymbol{x} = \boldsymbol{0}$. We seek $\boldsymbol{x} \neq \boldsymbol{0}$ that solves this homogeneous system. However, a square homogeneous system has a nontrivial solution if and only if its coefficient matrix is singular. Matrix $\boldsymbol{A}-\lambda\boldsymbol{I}$ is singular if and only if $\boldsymbol{det(A}-\lambda\boldsymbol{I)} = 0$. Thus eigenvalue λ is viewed as a tuning parameter to force matrix $\boldsymbol{A}-\lambda\boldsymbol{I}$ to be singular. It follows that with such values of λ we are able to determine nonzero vectors $\boldsymbol{x}$ so that $\boldsymbol{Ax} = \lambda\boldsymbol{x}$.

The expression $\boldsymbol{det(A}-\lambda\boldsymbol{I)}$ gives a polynomial of degree n in parameter λ which we call the <u>characteristic polynomial</u> of matrix $\boldsymbol{A}$. The expression $\boldsymbol{det(A}-\lambda\boldsymbol{I)} = 0$ is called the <u>characteristic equation</u> of matrix $\boldsymbol{A}$.

The eigenvalues of $\boldsymbol{A}$ are solutions (roots) of the characteristic equation. The corresponding eigenvectors are the solutions of the homogeneous system $(\boldsymbol{A}-\lambda\boldsymbol{I})\boldsymbol{x} = \boldsymbol{0}$.

Computationally we find the roots of the characteristic polynomial to determine the eigenvalues and then find the general solution of the corresponding homogeneous systems to find the eigenvectors. Hence the solution of the eigenproblem for matrix $\boldsymbol{A}$ is done in two steps.

In MATLAB , once matrix $\boldsymbol{A}$ is entered, we first find the characteristic polynomial of $\boldsymbol{A}$. Command

poly(A)

gives a vector containing the coefficients of the characteristic polynomial with the coefficient of the highest power term displayed first and the constant term last. (Zeros are used for the coefficient of any power of λ that is explicitly missing.) Command

roots(poly(A))

gives a vector containing the roots of the characteristic polynomial, that is the eigenvalues of $\boldsymbol{A}$. We illustrate these commands in the following example.

Example 1. Let $\boldsymbol{A} = \begin{bmatrix} 5 & -8 & -1 \\ 4 & -7 & -4 \\ 0 & 0 & 4 \end{bmatrix}$. Enter $\boldsymbol{A}$ into MATLAB . Then command

$$\mathbf{c = poly(A)}$$

displays

```
c =

     1    -2   -11    12
```

which implies that the characteristic polynomial of $\boldsymbol{A}$ is

$$1\lambda^3 - 2\lambda^2 - 11\lambda + 12$$

Using command

$$\mathbf{r = roots(poly(A))}$$

displays (in format short)

```
r =

   4.0000
  -3.0000
   1.0000
```

Note: If exact arithmetic had been used, then the roots of the characteristic polynomial for this matrix would have been integers 4, −3, and 1. Displaying **r** in format long e shows that a small amount of roundoff error occurred in the computation and hence MATLAB could not display the exact integer values. Such situations frequently occur in finding the roots of a characteristic polynomial in MATLAB (and other software).

The eigenvalues of $\boldsymbol{A}$ are $\lambda = 4, -3,$ and 1.

Once we have the eigenvalues λ of a matrix $\boldsymbol{A}$, the eigenvectors are determined as nontrivial solutions $\boldsymbol{x}$ of the homogeneous system $(\boldsymbol{A}-\lambda\boldsymbol{I})\boldsymbol{x} = \mathbf{0}$. To find $\boldsymbol{x} \neq \mathbf{0}$, compute rref$(\boldsymbol{A}-\lambda\boldsymbol{I})$ and construct the general solution of the homogeneous system. Linearly independent eigenvectors corresponding to λ are often obtained by extracting a basis for the general solution. This is equivalent to finding a basis for the null space of $\boldsymbol{A}-\lambda\boldsymbol{I}$ or a basis for the kernel of the linear transformation defined by $L(\boldsymbol{x}) = (\boldsymbol{A}-\lambda\boldsymbol{I})\boldsymbol{x}$. It is also a fact that eigenvectors corresponding to different eigenvalues are linearly independent. (An alternate approach uses routine **homsoln** applied to $\boldsymbol{A} - \lambda\boldsymbol{I}$ for each eigenvalue λ, which must be exactly specified.)

Example 2. For $\boldsymbol{A} = \begin{bmatrix} 5 & -8 & -1 \\ 4 & -7 & -4 \\ 0 & 0 & 4 \end{bmatrix}$, as defined in Example 1, the eigenvalues are $\lambda = 4, -3$, and 1. To find the corresponding eigenvectors in MATLAB proceed as follows.

Case $\lambda = 4$: MATLAB command

$$\mathbf{M} = \mathbf{rref(A} \ - \ 4{*}\mathbf{eye(size(A)))}$$

displays

```
M =

     1     0    -1
     0     1     0
     0     0     0
```

The general solution of $(\boldsymbol{A} - \mathbf{4}\boldsymbol{I})\boldsymbol{x} = \mathbf{0}$ is given by

$$x_3 = r, \qquad x_2 = 0, \qquad x_1 = r.$$

Hence $\boldsymbol{x} = r \begin{bmatrix} 1 \\ 0 \\ 1 \end{bmatrix}$ and we take $\begin{bmatrix} 1 \\ 0 \\ 1 \end{bmatrix}$ as an eigenvector corresponding to eigenvalue $\lambda = 4$. Note that we could have set constant r to any nonzero value to obtain an eigenvector. Hence eigenvectors corresponding to an eigenvalue are not unique.

Case $\lambda = -3$: MATLAB command

$$\mathbf{M} = \mathbf{rref(A} \ - \ (-3){*}\mathbf{eye(size(A)))}$$

displays

```
M =

     1    -1     0
     0     0     1
     0     0     0
```

The general solution of $(\boldsymbol{A} + \mathbf{3}\boldsymbol{I})\boldsymbol{x} = \mathbf{0}$ is given by

$$x_3 = 0, \qquad x_2 = r, \qquad x_1 = r$$

Hence $\boldsymbol{x} = r \begin{bmatrix} 1 \\ 1 \\ 0 \end{bmatrix}$ and we take $\begin{bmatrix} 1 \\ 1 \\ 0 \end{bmatrix}$ as an eigenvector corresponding to eigenvalue $\lambda = -3$.

Case $\lambda = 1$: MATLAB command

$$\mathbf{M = rref(A\ -\ 1{*}eye(size(A)))}$$

displays

```
M =

     1    -2     0
     0     0     1
     0     0     0
```

The general solution of $(\boldsymbol{A} - 1\boldsymbol{I})\boldsymbol{x} = \mathbf{0}$ is given by

$$x_3 = 0, \qquad x_2 = r, \qquad x_1 = 2r.$$

Hence $\boldsymbol{x} = r\begin{bmatrix} 2 \\ 1 \\ 0 \end{bmatrix}$ and we take $\begin{bmatrix} 2 \\ 1 \\ 0 \end{bmatrix}$ as an eigenvector corresponding to eigenvalue $\lambda = 1$.

Since matrix $\boldsymbol{A}$ had 3 distinct eigenvalues it follows that the eigenvectors $\begin{bmatrix} 1 \\ 0 \\ 1 \end{bmatrix}$, $\begin{bmatrix} 1 \\ 1 \\ 0 \end{bmatrix}$, and $\begin{bmatrix} 2 \\ 1 \\ 0 \end{bmatrix}$ are linearly independent.

Example 3. Let $\boldsymbol{A} = \begin{bmatrix} 7 & -4 & 0 \\ 8 & -5 & 0 \\ -4 & 4 & 3 \end{bmatrix}$. Command $\mathbf{r = roots(poly(A))}$ reveals that the eigenvalues of $\boldsymbol{A}$ are $\lambda = 3, 3, -1$. We find the eigenvectors corresponding to $\lambda = 3$ as follows. (We omit the case for $\lambda = -1$.)

Case $\lambda = 3$:

$$\mathbf{M = rref(A\ -\ 3{*}eye(size(A)))}$$

displays

```
M =

     1    -1     0
     0     0     0
     0     0     0
```

Hence $x_3 = r, x_2 = s, x_1 = s$ and we have

$$\boldsymbol{x} = \begin{bmatrix} s \\ s \\ r \end{bmatrix} = s \begin{bmatrix} 1 \\ 1 \\ 0 \end{bmatrix} + r \begin{bmatrix} 0 \\ 0 \\ 1 \end{bmatrix}.$$

It follows that both $\begin{bmatrix} 1 \\ 1 \\ 0 \end{bmatrix}$ and $\begin{bmatrix} 0 \\ 0 \\ 1 \end{bmatrix}$ are eigenvectors corresponding to $\lambda = 3$. Since r and s are arbitrary it follows that $\begin{bmatrix} 1 \\ 1 \\ 0 \end{bmatrix}$ and $\begin{bmatrix} 0 \\ 0 \\ 1 \end{bmatrix}$ are a pair of linearly independent eigenvectors corresponding to $\lambda = 3$. (Alternatively, **homsoln(A - 3*eye(size(A)))** produces the same eigenvectors.)

Warning: If a matrix has a k-times repeated eigenvalue, then it is possible that there will be fewer than k corresponding linearly independent eigenvectors. Such matrices are called **defective**. In a computing environment defective matrices may be difficult to recognize because of roundoff error within computations.

In MATLAB type

help eig

The display gives a description of the **eig** command. We are concerned only with the following features:

- **eig(A)** displays a vector containing the eigenvalues of square matrix $\boldsymbol{A}$.
- **[v,d] = eig(A)** displays the eigenvectors of $\boldsymbol{A}$ as columns of matrix $\boldsymbol{v}$ and the diagonal matrix $\boldsymbol{d}$ contains the corresponding eigenvalues.

We illustrate command **eig** in the following example.

Example 4. Enter matrix $\boldsymbol{A} = \begin{bmatrix} 3 & 0 & 0 \\ 4 & 2 & 1.5 \\ -5 & 0 & .5 \end{bmatrix}$ into MATLAB . Command

r = eig(A)

displays

```
r =

    2.0000
    0.5000
    3.0000
```

Command

$$[\mathbf{v},\mathbf{d}] = \mathbf{eig(A)}$$

displays

```
v =                                  d =
        0         0    0.4082            2.0000         0         0
   1.0000    -.7071    0.4082                 0    0.5000         0
        0     .7071   -0.8165                 0         0    3.0000
```

The columns of $\boldsymbol{v}$ are the eigenvectors of $\boldsymbol{A}$ corresponding to the eigenvalues in the diagonal entries of the same numbered column of $\boldsymbol{d}$. It is MATLAB 's convention that the eigenvectors are scaled (multiplied by a nonzero scalar) so that their norm is 1. Had we done the computations by hand the matrix $\boldsymbol{v}$ could have been displayed as

```
0   0   1
1  -1   1
0   1  -2
```

Warning: MATLAB computes eigenvalues and eigenvectors by methods different from those we have studied. The results are quite accurate, but may appear different from the corresponding hand calculations.

Exercises 13.2

1. Use commands **poly** and **roots** to find the characteristic polynomial and the eigenvalues of each of the following matrices. Record your results below each matrix.

a) $\boldsymbol{A} = \begin{bmatrix} 4 & -2 & -5 \\ 1 & 1 & -1 \\ 0 & 0 & -1 \end{bmatrix}$

b) $\boldsymbol{B} = \begin{bmatrix} -6 & 8 & 1 \\ -4 & 6 & 1 \\ 0 & 0 & 1 \end{bmatrix}$

c) $C = \begin{bmatrix} -1/2 & 1 & -1/2 \\ -1/2 & 1 & -1/2 \\ 0 & 0 & 1 \end{bmatrix}$

d) $D = \begin{bmatrix} 1 & 2 & 0 & 0 \\ 2 & 1 & 0 & 0 \\ 0 & 0 & 1 & 1 \\ 0 & 0 & 1 & 1 \end{bmatrix}$

______________________ ______________________

______________________ ______________________

2. Here we investigate the eigenvalues of upper triangular matrices. Perform the following experiments. Record the matrix $\boldsymbol{A}$ and its eigenvalues. Look for a connection between the entries of $\boldsymbol{A}$ and its eigenvalues.

n = 3; A = triu(fix(10*rand(n))), r = roots(poly(A))

Repeat the experiment several times. Then change n to 4 and do the experiment. Complete the following conjecture:

The eigenvalues of an upper triangular matrix are ______________________.

Check your conjecture by changing n to 5 and repeat the experiment several times.

3. In the MATLAB commands in Exercise 2, replace **triu** by **tril** and investigate the eigenvalues of a lower triangular matrix. Look for a connection between the entries of $\boldsymbol{A}$ and its eigenvalues. Complete the following conjecture:

The eigenvalues of a lower triangular matrix are ______________________.

4. Using the results from Exercises 2 and 3 fill in the following conjecture.

The eigenvalues of a diagonal matrix are ______________________.

Give a reason for this conjecture based on ideas relating triangular and diagonal matrices.

____

5. Find the eigenvalues and eigenvectors for each of the following. Record your results below each matrix.

a) $A = \begin{bmatrix} 3 & 0 & -1 \\ -1 & -6 & 9 \\ -1 & 0 & 3 \end{bmatrix}$ b) $B = \begin{bmatrix} 0 & 1 & 0 \\ 0 & -1 & 0 \\ -2 & -2 & -1 \end{bmatrix}$

6. Let $A = \begin{bmatrix} 0 & 1 \\ -1 & 0 \end{bmatrix}$. Find the eigenvalues of A. (The eigenvalues will be complex even though the matrix A is real.) Find the eigenvectors as in Example 2. Record your results below.

7. Find the eigenvalues and eigenvectors of complex matrix $A = \begin{bmatrix} 3-2i & -1-2i & 0 \\ 0 & 4 & 0 \\ -2+6i & 5i & 2+i \end{bmatrix}$ using the methods in Examples 1 and 2. Record your results below. (Note that A has complex eigenvalues, but its eigenvectors are real.)

8. Use **eig** on the matrices in Exercise 1. Check the eigenvalues with those you computed using **poly** and **roots**.

9. Use **eig** on the matrices in Exercise 5. Check your results from Exercise 5.

10. Do Exercise 7 using **eig**.

Section 13.3

Further Experiments and Applications

Following are a collection of experiments on eigenvalues and eigenvectors that provide opportunities to study properties of these concepts. We also provide an introduction to diagonalizable matrices using MATLAB experiments. Applications involving powers of a matrix, Markov matrices, and graph theory are explored briefly.

Exercises 13.3

1. Let $\boldsymbol{A} = \begin{bmatrix} -11 & 0 & -1 & 13 \\ 12 & 1 & 1 & -13 \\ 10 & 0 & 4 & -10 \\ -5 & 0 & -1 & 7 \end{bmatrix}$. Compute the eigenvalues of $\boldsymbol{A}$, $\boldsymbol{A}^2$, and $\boldsymbol{A}^3$. Record your results below.

By inspecting the results above, complete the following conjectures.

The eigenvalues of $\boldsymbol{A}^2$ are the ____________ of the eigenvalues of $\boldsymbol{A}$.

The eigenvalues of $\boldsymbol{A}^3$ are the ____________ of the eigenvalues of $\boldsymbol{A}$.

Check to see if the pattern of your conjectures above continues by computing the eigenvalues of $\boldsymbol{A}^4$ and $\boldsymbol{A}^5$. Change the matrix $\boldsymbol{A}$ and repeat this series of experiments.

Complete the following conjecture:

If λ is an eigenvalue of $\boldsymbol{A}$, then ____________ is an eigenvalue of $\boldsymbol{A}^k$.

2. Let $\boldsymbol{A} = \begin{bmatrix} 8 & 0 & 11 \\ 5.5 & 1.8 & 10.3 \\ -5.5 & 0 & -8.5 \end{bmatrix}$. Compute and record the eigenvalues of $\boldsymbol{A}$.

____

Perform and record the results of the following calculations.

Find the eigenvalues of $\boldsymbol{A}+3\boldsymbol{I}$ ____________
Find the eigenvalues of $\boldsymbol{A}-\boldsymbol{I}$ ____________
Find the eigenvalues of $\boldsymbol{A}+2\boldsymbol{I}$ ____________

Formulate a conjecture concerning the relationship between the eigenvalues of $\boldsymbol{A}$ and those of $\boldsymbol{A}+k\boldsymbol{I}$ by completing the following.

If λ is an eigenvalue of $\boldsymbol{A}$, then ____________ is an eigenvalue of $\boldsymbol{A}+k\boldsymbol{I}$.

3. If $\boldsymbol{A}$ is $n \times n$ and has n linearly independent eigenvectors which are used to form matrix $\boldsymbol{P}$, then

$$\boldsymbol{P}^{-1}\boldsymbol{A}\boldsymbol{P} = \boldsymbol{D}$$

where $\boldsymbol{D}$ is a diagonal matrix. The diagonal entries of $\boldsymbol{D}$ are the eigenvalues corresponding to the eigenvectors which are the columns of $\boldsymbol{P}$. Matrix $\boldsymbol{A}$ is called diagonalizable or we say that $\boldsymbol{A}$ is similar to a diagonal matrix. For such matrices $\boldsymbol{A}$ we have

$$\boldsymbol{A} = \boldsymbol{P}\boldsymbol{D}\boldsymbol{P}^{-1}$$

The powers of $\boldsymbol{A}$ are easily expressed in terms of the powers of the diagonal matrix $\boldsymbol{D}$ as follows:

$$\boldsymbol{A}^2 = (\boldsymbol{PDP}^{-1})\,(\boldsymbol{PDP}^{-1}) = \boldsymbol{PDP}^{-1}\boldsymbol{PDP}^{-1} = \boldsymbol{PDIDP}^{-1} = \boldsymbol{PD}^2\boldsymbol{P}^{-1}$$

$$\boldsymbol{A}^3 = \boldsymbol{AA}^2 = (\boldsymbol{PDP}^{-1})(\boldsymbol{PD}^2\boldsymbol{P}^{-1}) = \boldsymbol{PD}^3\boldsymbol{P}^{-1}$$

and in general $\boldsymbol{A}^k = \boldsymbol{PD}^k\boldsymbol{P}^{-1}$

Next note that for diagonal matrix

$$\boldsymbol{D} = \begin{bmatrix} d_1 & 0 & 0 & \cdots & 0 \\ 0 & d_2 & 0 & \cdots & 0 \\ . & . & . & . & . \\ . & . & . & . & . \\ 0 & 0 & \cdots & 0 & d_n \end{bmatrix}$$

it follows that

$$\boldsymbol{D}^k = \begin{bmatrix} (d_1)^k & 0 & 0 & \cdots & 0 \\ 0 & (d_2)^k & 0 & \cdots & 0 \\ . & . & . & . & . \\ . & . & . & . & . \\ 0 & 0 & \cdots & 0 & (d_n)^k \end{bmatrix}.$$

Hence in MATLAB we can compute $\boldsymbol{A}^k$ much more economically than by repeated multiplication as $\boldsymbol{A} * \boldsymbol{A} * \boldsymbol{A} * \cdots * \boldsymbol{A}$. Consider the following strategy.

- Find the eigenvalues and eigenvectors of $\boldsymbol{A}$ using command **[P,D] = eig(A)**.
- Find $\boldsymbol{P}^{-1}$ using the **inv** command; **Pinv = inv(P);**.
- Set **s = diag(D)**. Then vector **s** contains the eigenvalues of $\boldsymbol{A}$. Next raise each eigenvalue to the kth power using **sk = s.^k** with the desired value of k.
- Compute $\boldsymbol{A}^k$ using **P * diag(sk) * Pinv**.
- Summary: These steps can be combined into the single MATLAB command line

  $$\boxed{\mathbf{[P,D] = eig(A); Q = P * (diag(diag(D).^\wedge k)) * inv(P)}}$$

  Replace k by the desired value and then matrix $\boldsymbol{Q}$ is $\boldsymbol{A}^k$.

To compare the work involved in computing $\boldsymbol{A}^k$ directly by repeated multiplication and by the eigenmethod as outlined above for a symmetric matrix use the following experiment.

(Here **flops** is a command that counts the number of arithmetic operations used. For more information type **help flops**).

n = 8;A = rand(n);	(setting matrix size and generating a matrix)
A = triu(A)+triu(A,1)';	(constructing a symmetric matrix)
flops(0);	(setting flop count to zero)
k = 20;B = A^k;	(setting power and doing direct computation)
df = flops	(displaying flop count for direct computation)
flops(0);	(resetting flop count to zero)
[P,D] = eig(A);	
C = P*(diag(diag(D).^k))*inv(P);	
idf = flops	(displaying flop count for indirect method)

A measure of the difference in the work involved is obtained by comparing the values of df, the number of flops for direct computation, and idf, the number of flops for indirect computation.

a) Repeat the experiment for a 15×15 matrix raised to the 20th power. (n = 15 and k = 20)

df = ____________ idf = ____________

b) Repeat the experiment for a 25×25 matrix raised to the 25th power.

df = ____________ idf = ____________

c) Repeat the experiment for a 25×25 matrix raised to the 50th power.

df = ____________ idf = ____________

4. A wide variety of applications depend upon the behavior of the powers of a square matrix $\boldsymbol{A}$. For instance, the long term behavior of the population model that we studied in Lab 3 required us to predict future populations using computations like

$$\boldsymbol{Ax},\ \boldsymbol{A}^2\boldsymbol{x},\ \boldsymbol{A}^3\boldsymbol{x},\ \ldots$$

where $\boldsymbol{x}$ represents an initial population. The long term behavior of this model really depends on

$$\lim_{k\to\infty} \boldsymbol{A}^k.$$

If $\boldsymbol{A}$ is similar to a diagonal matrix then $\boldsymbol{A} = \boldsymbol{PDP}^{-1}$ and it follows that

$$\lim_{k\to\infty} \boldsymbol{A^k} = \lim_{k\to\infty} (\boldsymbol{PDP}^{-1})^k = \boldsymbol{P}(\lim_{k\to\infty} \boldsymbol{D^k})\boldsymbol{P}^{-1}.$$

Since

$$\boldsymbol{D^k} = \begin{bmatrix} (d_1)^k & 0 & 0 & \cdots & 0 \\ 0 & (d_2)^k & 0 & \cdots & 0 \\ \cdot & \cdot & \cdot & \cdot & \cdot \\ \cdot & \cdot & \cdot & \cdot & \cdot \\ 0 & 0 & \cdots & 0 & (d_n)^k \end{bmatrix}.$$

$\lim_{k\to\infty} D^k$ depends on the $\lim_{k\to\infty} d_j^k$ for $j = 1, 2, \ldots, n$. That is, we merely investigate the behavior of the powers of the eigenvalues of $\boldsymbol{A}$. Hence the behavior of the sequence $\boldsymbol{Ax}$, $\boldsymbol{A}^2\boldsymbol{x}$, $\boldsymbol{A}^3\boldsymbol{x}$, ... is determined by the behavior of the sequences of powers of the eigenvalues of $\boldsymbol{A}$. We simplify the analysis even further as follows. Let

$$\mathbf{r} = \{\mathbf{max} \mid \lambda \mid\}$$

where λ is an eigenvalue of $\boldsymbol{A}$. Then the behavior of sequence $\boldsymbol{Ax}$, $\boldsymbol{A}^2\boldsymbol{x}$, $\boldsymbol{A}^3\boldsymbol{x}$, ... for any initial vector $\boldsymbol{x}$ is determined by the behavior of

$$\lim_{k\to\infty} r^k.$$

We have three cases:

- If $r < 1$, then $\lim_{k\to\infty} \boldsymbol{A^k x} = \boldsymbol{0}$ and we say that the sequence $\boldsymbol{Ax}$, $\boldsymbol{A}^2\boldsymbol{x}$, $\boldsymbol{A}^3\boldsymbol{x}$, ... is stable.
- If $r = 1$, then $\mathbf{lim}_{k\to\infty} \boldsymbol{A^k x} = \boldsymbol{u}$ where $\boldsymbol{u} \neq \boldsymbol{0}$ and we say that the sequence $\boldsymbol{Ax}$, $\boldsymbol{A}^2\boldsymbol{x}$, $\boldsymbol{A}^3\boldsymbol{x}$, ... is neutrally stable[1]. Alternatively, we say the process has a nontrivial steady state.
- If $r > 1$, then $\mathbf{lim}_{k\to\infty} \boldsymbol{A^k x}$ does not exist and we say that the sequence $\boldsymbol{Ax}$, $\boldsymbol{A}^2\boldsymbol{x}$, $\boldsymbol{A}^3\boldsymbol{x}$, ... is unstable.

[1]We assume that the only eigenvalue with magnitude one is $\lambda = 1$, which is the case for most population matrices.

a) If sequence $\boldsymbol{Ax}$, $\boldsymbol{A}^2\boldsymbol{x}$, $\boldsymbol{A}^3\boldsymbol{x}$, ... represents populations from year to year, then write a brief description about the population behavior for each of the following cases:

stable ________________

neutrally stable ________________

unstable ________________

b) For each of the following population transition matrices $\boldsymbol{A}$ determine if the process defined by the sequence $\boldsymbol{Ax}$, $\boldsymbol{A}^2\boldsymbol{x}$, $\boldsymbol{A}^3\boldsymbol{x}$, ... is stable, neutrally stable, or unstable. Record your findings next to matrix.

i) $\boldsymbol{A} = \begin{bmatrix} .559 & .6 & .1 \\ .7 & 0 & 0 \\ 0 & .3 & 0 \end{bmatrix}$ ________________

ii) $\boldsymbol{A} = \begin{bmatrix} .21 & .64 & .12 \\ .69 & 0 & 0 \\ 0 & .36 & 0 \end{bmatrix}$ ________________

iii) $\boldsymbol{A} = \begin{bmatrix} .868 & .4 & .2 \\ .3 & 0 & 0 \\ 0 & .2 & 0 \end{bmatrix}$ ________________

c) For the population transition matrix $\boldsymbol{A} = \begin{bmatrix} .5 & .44 & .06 \\ 1 & 0 & 0 \\ 0 & 1 & 0 \end{bmatrix}$ it was determined directly that for initial population vector $\boldsymbol{x} = \begin{bmatrix} 14 \\ 20 \\ 11 \end{bmatrix}$ (in millions) that

$$\lim_{k\to\infty} \mathbf{A}^k\mathbf{x} = \mathbf{u} = \begin{bmatrix} 15.8077 \\ 15.8077 \\ 15.8077 \end{bmatrix}.$$

How is $\boldsymbol{u}$ related to the eigenvectors of $\boldsymbol{A}$? Be specific. ________________

5. A **Markov** matrix $\boldsymbol{M}$ has each $m_{ij} \geq 0$ and the sum of each column equal to 1. For a given matrix $\boldsymbol{M}$ we can inspect its entries to determine if they are nonnegative and MATLAB can be used to compute the sum of the columns. Type **help sum** for a description of the behavior of command **sum(M)**.

$$M1 = \begin{bmatrix} .1 & .3 & .4 \\ .5 & .2 & .4 \\ .4 & .3 & .2 \end{bmatrix} \qquad M2 = \begin{bmatrix} .2 & .2 & .6 \\ .5 & .2 & .4 \\ .3 & .6 & .1 \end{bmatrix}$$

$$M3 = \begin{bmatrix} .5 & .1 & .6 & .4 \\ .2 & .5 & .2 & .1 \\ .1 & .3 & 0 & .1 \\ .2 & .1 & .2 & .4 \end{bmatrix} \qquad M4 = \begin{bmatrix} .3 & .21 & .46 & .4 \\ .2 & .59 & .24 & .1 \\ .3 & .2 & .19 & .1 \\ .2 & .1 & .11 & .4 \end{bmatrix}$$

a) Use MATLAB to determine which of $M1$ through $M4$ is a Markov matrix. (Indicate Yes or No.)

$M1$ __________ $M2$ __________ $M3$ __________ $M4$ __________

b) For those that are Markov matrices use **eig** to determine $r = \{max \mid \lambda \mid\}$ where λ is an eigenvalue. Record your results here.

c) Complete the following using the terminology of Exercise 4.

Any process using a Markov matrix is __________.

Check your conjecture on a 4×4 Markov matrix that you construct.

d) For each of the Markov matrices of $M1$ through $M4$ find the eigenvector corresponding to r from part b. How are these eigenvectors related to your conjecture in part c? Be specific.

6. Lab 3 investigated the populations of three age groups of an animal species that lived to a maximum of two years, using the transition matrix

$$A = \begin{bmatrix} .559 & .6 & .1 \\ .7 & 0 & 0 \\ 0 & .3 & 0 \end{bmatrix}$$

The problem of determining the stable distribution for each age group can best be solved using eigenvalues and eigenvectors.

a) What are the eigenvalues of $\boldsymbol{A}$? ____________

b) Determine an eigenvector $\boldsymbol{v}$ corresponding to eigenvalue $\lambda = 1$. ____________

c) Determine an eigenvector $\boldsymbol{v}$ corresponding to eigenvalue $\lambda = 1$ that has positive entries. ____________

d) Determine an eigenvector $\boldsymbol{v}$ corresponding to eigenvalue $\lambda = 1$ that has positive entries and the sum of the entries is equal to 1. ____________

e) Use the vector in part d) to determine the percentage of animal species that will eventually have

current age zero. ____________

current age one. ____________

current age two. ____________

7. Graph theory is a valuable modeling tool. In graph theory a **graph** G provides a model of set of objects and their relationships. The objects involved are represented as points $P_1, P_2, \ldots P_k$ and their relationships are denoted by directed paths between points. The points are called **nodes** or **vertices** and the paths are called **edges**. For example, if the set of trade routes between four cities $P_1, P_2, P_3,$ and P_4 are represented by Figure 1 where each edge can be traversed in either direction. Next we define the **incidence matrix** $\boldsymbol{A}$ of such a graph by

$$a_{ij} = \begin{cases} 1, \text{ if there is a route between } P_i \text{ and } P_j,\ i \neq j \\ 0, \text{ otherwise} \end{cases}$$

For the graph in Figure 1, the adjacency matrix is

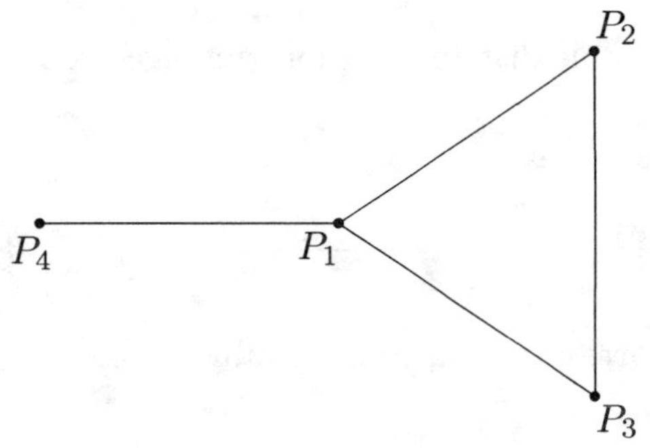

Figure 1

$$A = \begin{bmatrix} 0 & 1 & 1 & 1 \\ 1 & 0 & 1 & 0 \\ 1 & 1 & 0 & 0 \\ 1 & 0 & 0 & 0 \end{bmatrix}$$

Note that $\boldsymbol{A}$ is symmetric. (For other applications incidence matrices are defined differently and need not be symmetric.)

a) By inspecting the graph, rank each city in the trade route network by its accessibility to the others. The highest rank to the most accessible. Use 5 for the highest with cities of equal accessibility assigned the same value.

City P_1 ____________
City P_2 ____________
City P_3 ____________
City P_4 ____________

b) Compute the eigenvalues of $\boldsymbol{A}$. Record the results here.

____

c) Compute the eigenvector $\mathbf{v}$ corresponding to the largest eigenvalue and define $\mathbf{u} = \mathbf{v}/\mathbf{norm}(\mathbf{v})$. Record $\mathbf{u}$ here.

____

d) Consider the components of **u** as a measure of the accessibility of cities 1, 2, 3, and 4 respectively. Do these agree, in principal, with your ranking in part a)? Explain.

8. Repeat the analysis outlined in Exercise 7 on the graph in Figure 2.

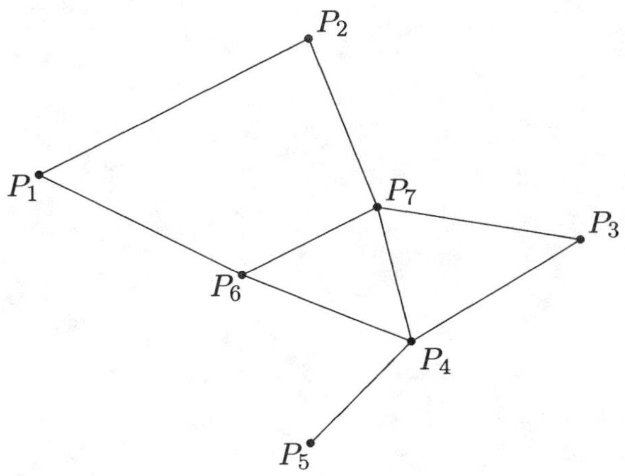

Figure 2

<< NOTES; COMMENTS; IDEAS >>

<< NOTES; COMMENTS; IDEAS >>

Projects

- Introduction To Graph Theory
- Secret Codes
- Least Squares Models

<< **NOTES; COMMENTS; IDEAS** >>

Introduction to Graph Theory

Description: We introduce the notion of a graph to represent information. This information is then presented mathematically using the notion of an incidence matrix. We develop the computations for determining the number of paths of various lengths connecting nodes of the graph. Digraphs are discussed in the exercises.

Prerequisites: matrix sums, products, and powers; symmetric matrices.

PROJECT 1

Incidence Matrices

By a **graph** we mean a set of points some of which are connected by line segments. The points are often called **nodes** or **vertices** and the line segments are called **edges**. Each of the following figures is a graph.

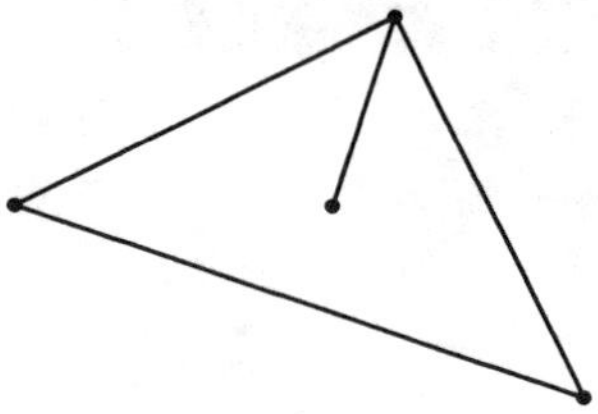

Figure 1

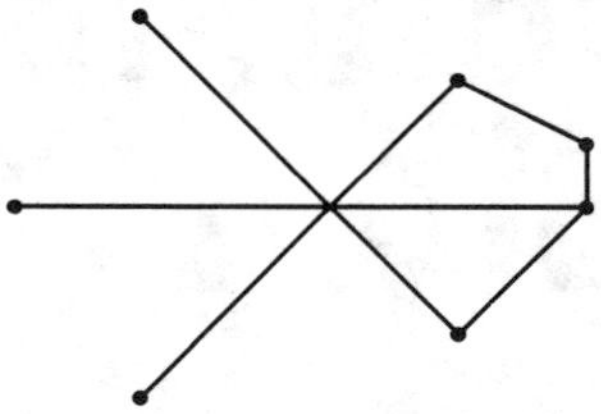

Figure 2

The nodes are usually labeled as P1, P2, ... and for now we allow an edge to be traversed in either direction. See Figure 3.

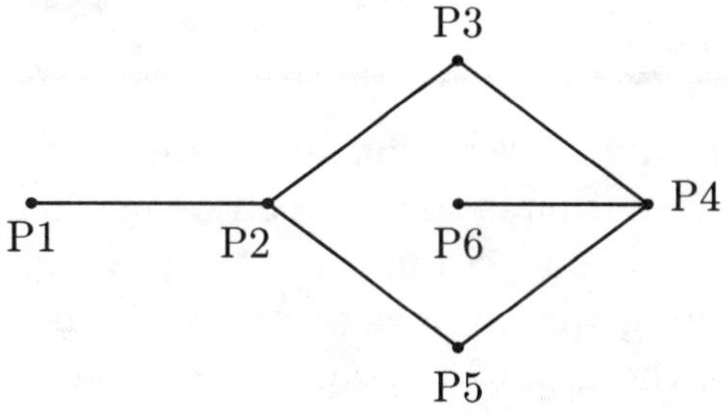

Figure 3

Graphs are becoming increasingly important to represent information. Certainly transportation networks by rail, road, or air have a natural representation as a graph. But so do telephone and information networks, designs for electric circuitry, the study of social structures, analysis of ancient trade routes, molecular models, and an ever increasing number of other topics.

One way to represent a graph mathematically is by an **incidence matrix *A***. Each row and corresponding column of the incidence matrix ***A*** is associated with a node. A natural way to assign the nodes is to associate row j and column j with node Pj. Then the incidence matrix is given entries as follows:

$$a_{ij} = \begin{cases} 0, & \text{if there is no edge connecting Pi and Pj} \\ 1, & \text{if there is an edge connecting Pi and Pj} \end{cases}$$

We illustrate this in examples.

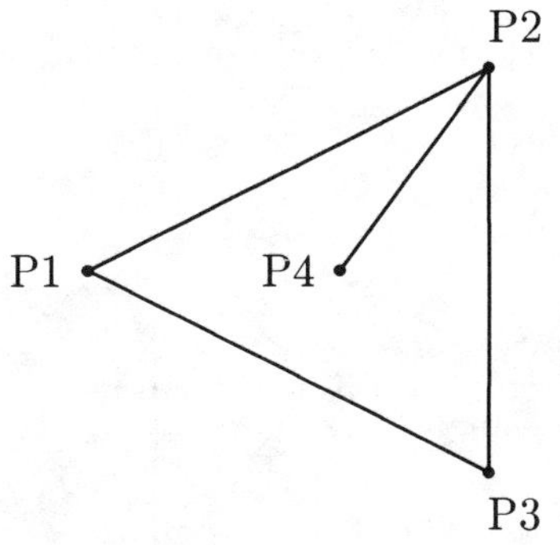

Figure 4

Example 1. Suppose we label Figure 1 and rename it Figure 4 as shown.
The incidence matrix associated with the graph in Figure 4 is constructed as follows.

$\boldsymbol{A} =$

	P1	P2	P3	P4
P1				
P2				
P3				
P4				

where

$a_{11} = 0$ since there is no edge from P1 to P1.
$a_{12} = 1$ since there is an edge from P1 to P2.
Note also that $a_{21} = 1$ since the edge from P1 to P2 can be traversed from P2 to P1.
$a_{14} = 0$ since there is no edge from P1 to P4. Similarly, $a_{41} = 0$.
etc.

Verify that the full incidence matrix $\boldsymbol{A}$ for Figure 4 is

$$\boldsymbol{A} = \begin{bmatrix} 0 & 1 & 1 & 0 \\ 1 & 0 & 1 & 1 \\ 1 & 1 & 0 & 0 \\ 0 & 1 & 0 & 0 \end{bmatrix}$$

Note that matrix $\boldsymbol{A}$ is symmetric; that is, $\boldsymbol{A}^T = \boldsymbol{A}$. This follows since an edge can be traversed in either direction. (If we permit 'one-way' edges the incidence matrix need not be symmetric. See Exercises 3 to 7.)

Example 2. In Figure 4 suppose that a new edge is inserted from P4 to P3. Then the graph is

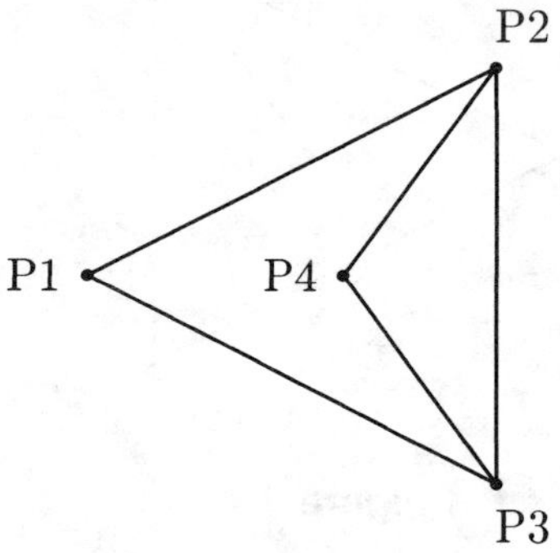

Figure 5

Construct the incidence matrix for this graph in the space below and name it $\boldsymbol{B}$.

$$\boldsymbol{B} = \begin{bmatrix} \\ \\ \\ \\ \end{bmatrix}$$

To check your incidence matrix $\boldsymbol{B}$ we can have MATLAB draw the graph associated with $\boldsymbol{B}$. Enter $\boldsymbol{B}$ into MATLAB and then enter command

igraph(B)

Does the display look like Figure 5? Write a short comparison in the space below. (Warning: It may be oriented differently.)

By hand draw the graph associated with incidence matrix

$$\boldsymbol{C} = \begin{bmatrix} 0 & 1 & 0 & 0 \\ 1 & 0 & 1 & 0 \\ 0 & 1 & 0 & 1 \\ 0 & 0 & 1 & 0 \end{bmatrix}$$ Put your graph here $\longrightarrow$

Check your graph using routine **igraph** in MATLAB . (For more information on routine **igraph** type **help igraph**.)

Paths

In a variety of applications the graph representing the information under consideration is used to determine a 'connection' or interrelationship between the nodes. The graph itself provides a visual picture of which pairs of nodes are connected by single edges and this information is directly reflected in the incidence matrix. Not every node is connected to every other node by an edge, so to make a connection between some nodes we must traverse more than one edge. It is convenient to use the term **path** to represent a connection between nodes and we can talk about paths of various lengths. (Here length just refers to the number of edges traversed.) We use the following conventions:

A path of length 1 traverses a single edge.
A path of length 2 traverses a pair of edges.
  Note: if there is a path of length 1 from
  P1 to P2, then there is a path of length 2 from
  P1 to P1 by using the route P1 to P2 to P1.
A path of length 3 traverses three edges.
etc.

In Figure 5 we have the following examples of paths:

There is a path of length 1 from P1 to P2.
There is a path of length 2 from P1 to P3;
  use route P1 to P2 to P3.
There are two paths of length 2 from P1 to P1;
  P1 to P2 to P1 and P1 to P3 to P1.

To set up a delivery schedule for a transportation network it is important to know how many paths of various lengths go from one node to another. This is also true for many other situations modeled by graphs. For instance, in Figure 5 count the number of paths of length 3 from P4 to P1. Careful inspection shows there are two:

P4 to P2 to P3 to P1
P4 to P3 to P2 to P1

Such counting by inspection is risky and very prone to error. To illustrate, try counting the paths of length 3 from P1 to P3 in Figure 5. There are five. One is P1 to P2 to P4 to P3; another is P1 to P3 to P4 to P3; another is P1 to P2 to P1 to P3. Find the other two and list them below.

For efficiency and accuracy we need a mechanism for counting the number of paths of a given length between pairs of nodes that does not rely on mere observation. This is where

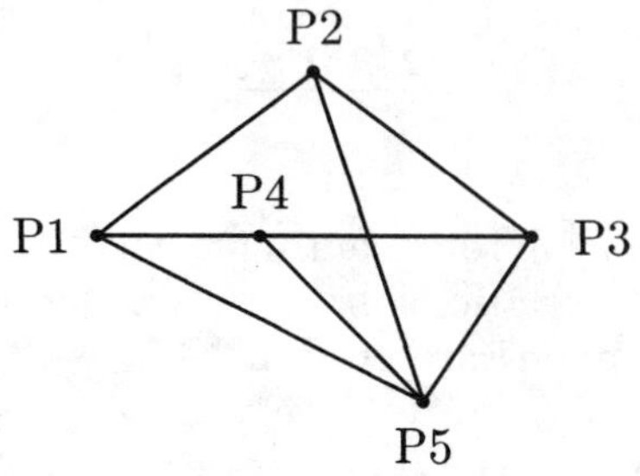

Figure 6

matrix operations can be used efficiently on an incidence matrix. We illustrate the case of paths of length 2 from P1 to P5 in Figure 6, whose incidence matrix E is displayed below it.

$$\boldsymbol{E} = \begin{bmatrix} 0 & 1 & 0 & 1 & 1 \\ 1 & 0 & 1 & 0 & 1 \\ 0 & 1 & 0 & 1 & 1 \\ 1 & 0 & 1 & 0 & 1 \\ 1 & 1 & 1 & 1 & 0 \end{bmatrix}$$

A path of length 2 from P1 to P5 must travel through another node Pj. Hence we must have a route of the form

P1 to Pj to P5

We first find the number of paths of length 1 from P1 to Pj for $j = 1, 2, 3, 4, 5$. These are respectively the entries in $\text{row}_1(\boldsymbol{E})$; namely $\begin{bmatrix} 0 & 1 & 0 & 1 & 1 \end{bmatrix}$. Next we find the number of paths of length 1 from Pj to P5. These are respectively the entries in $\text{col}_5(\boldsymbol{E})$; namely $\begin{bmatrix} 1 \\ 1 \\ 1 \\ 1 \\ 0 \end{bmatrix}$. We combine this information to count paths from P1 to Pj to P5 as follows: (We use symbol # to represent the word "number".)

Case $j = 1$:

$$\begin{matrix}\text{\# of paths from} \\ \text{P1 to P1 to P5}\end{matrix} = \left(\begin{matrix}\text{\# of paths from} \\ \text{P1 to P1}\end{matrix}\right) * \left(\begin{matrix}\text{\# of paths from} \\ \text{P1 to P5}\end{matrix}\right) = (a_{11} * a_{15}) = 0 * 1 = 0$$

Case $j = 2$:

$$\begin{matrix}\text{\# of paths from} \\ \text{P1 to P2 to P5}\end{matrix} = \left(\begin{matrix}\text{\# of paths from} \\ \text{P1 to P2}\end{matrix}\right) * \left(\begin{matrix}\text{\# of paths from} \\ \text{P2 to P5}\end{matrix}\right) = (a_{12} * a_{25}) = 1 * 1 = 1$$

Case $j = 3$:

$$\begin{matrix}\text{\# of paths from} \\ \text{P1 to P3 to P5}\end{matrix} = \begin{pmatrix}\text{\# of paths from} \\ \text{P1 to P3}\end{pmatrix} * \begin{pmatrix}\text{\# of paths from} \\ \text{P3 to P5}\end{pmatrix} = (a_{13} * a_{35}) = 0 * 1 = 0$$

Case $j = 4$:

$$\begin{matrix}\text{\# of paths from} \\ \text{P1 to P4 to P5}\end{matrix} = \begin{pmatrix}\text{\# of paths from} \\ \text{P1 to P4}\end{pmatrix} * \begin{pmatrix}\text{\# of paths from} \\ \text{P4 to P5}\end{pmatrix} = (a_{14} * a_{45}) = 1 * 1 = 1$$

Case $j = 5$:

$$\begin{matrix}\text{\# of paths from} \\ \text{P1 to P5 to P5}\end{matrix} = \begin{pmatrix}\text{\# of paths from} \\ \text{P1 to P5}\end{pmatrix} * \begin{pmatrix}\text{\# of paths from} \\ \text{P5 to P5}\end{pmatrix} = (a_{15} * a_{55}) = 1 * 0 = 0$$

The total number of paths of length 2 from P1 to P5 is the sum of the preceding quantities:

$$a_{11} * a_{15} + a_{12} * a_{25} + a_{13} * a_{35} + a_{14} * a_{45} + a_{15} * a_{55} = \sum_{j=1}^{5} a_{1j} * a_{j5} = row_1(\boldsymbol{E}) * col_5(\boldsymbol{E}) = 2$$

This computation generalizes in a natural way so that

$$\boxed{\text{\# of paths of length 2 from Pi to Pj} = \text{row}_i(\boldsymbol{E}) * \text{col}_j(\boldsymbol{E})}$$

But row-times-column is how we compute a matrix product. Hence it follows that the entries of $\boldsymbol{E}^2$ give the number of paths of length 2 between pairs of nodes:

$$\boxed{(\boldsymbol{E}^2)_{ij} = \text{\# of paths of length 2 from Pi to Pj}}$$

For the graph in Figure 6

$$\boldsymbol{E}^2 = \begin{bmatrix} 3 & 1 & 3 & 1 & 2 \\ 1 & 3 & 1 & 3 & 2 \\ 3 & 1 & 3 & 1 & 2 \\ 1 & 3 & 1 & 3 & 2 \\ 2 & 2 & 2 & 2 & 4 \end{bmatrix}$$

Thus $(\boldsymbol{E}^2)_{24} = 3$ implies that there are 3 paths of length 2 from P2 to P4.

We summarize the preceding development and extend it to paths of arbitrary length by the following statement.

> If $\boldsymbol{A}$ is an incidence matrix of a graph, then the number of paths of length k from Pi to Pj is the (i,j)-entry of $\boldsymbol{A}^k$.

With the characterization of the number of paths of length k between nodes in terms of the entries of the incidence matrix raised to the k-th power, we have a mechanism for checking effects of adding or deleting paths from the graph. For example, if the edge from P4 to P5 is deleted from Figure 6 we have Figure 7, whose incidence matrix is $\boldsymbol{F}$.

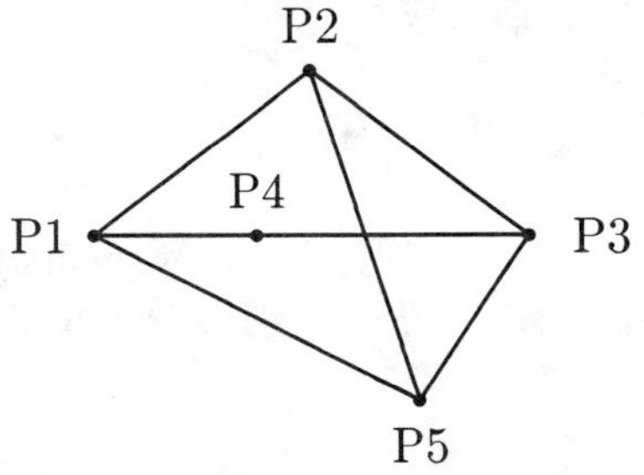

Figure 7

$$\boldsymbol{F} = \begin{bmatrix} 0 & 1 & 0 & 1 & 1 \\ 1 & 0 & 1 & 0 & 1 \\ 0 & 1 & 0 & 1 & 1 \\ 1 & 0 & 1 & 0 & 0 \\ 1 & 1 & 1 & 0 & 0 \end{bmatrix}$$

Determine the number of paths of length 2 from P1 to P5. Record your result. __________

Certain applications require that we know the longest path required between any pair of nodes in the graph. The entries of the incidence matrix $\boldsymbol{A}$ give the number of paths of length 1 between nodes, while the entries of $\boldsymbol{A}^2$ give the number of paths of length 2 between nodes. It follows that the (i,j)-entry of $\boldsymbol{A} + \boldsymbol{A}^2$ is the number of paths of length 2 or less from Pi to Pj. Similarly the (i,j)-entry of $\boldsymbol{A} + \boldsymbol{A}^2 + \boldsymbol{A}^3$ is the number of paths of length 3 or less from Pi to Pj. To determine the longest path required between any pair of nodes in the graph whose incidence matrix is $\boldsymbol{A}$, find the smallest value of k so that

$$\boldsymbol{A} + \boldsymbol{A}^2 + \cdots + \boldsymbol{A}^k$$

has all non-diagonal entries different from zero. This value k is the length of the longest path required between nodes in the graph. Show that k = 2 for the graph in Figure 7.

What is the value of k for the graph in Figure 8? __________

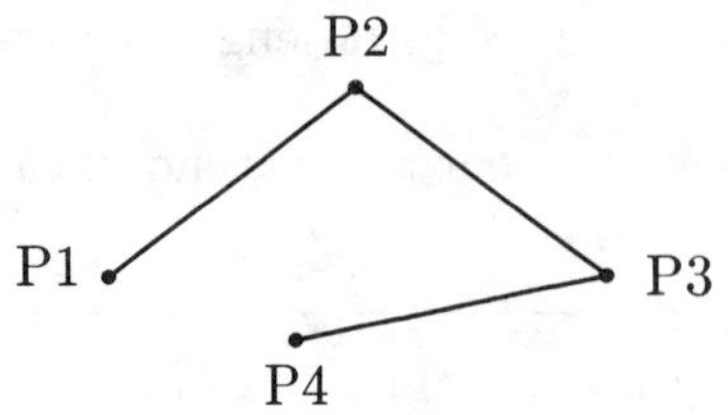

Figure 8

Exercises

1. Consider the graph in Figure 9.

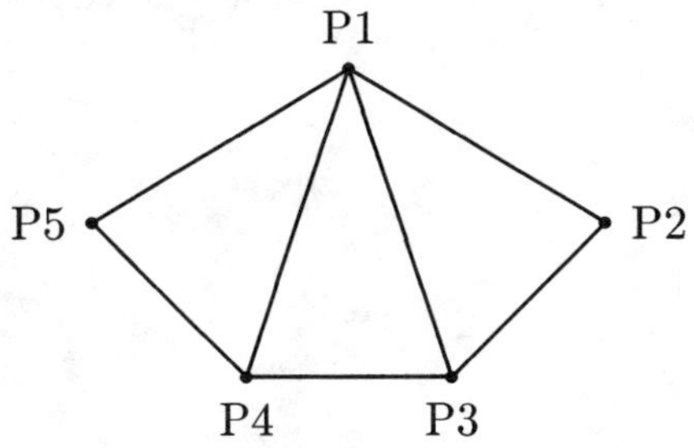

Figure 9

a) Construct the incidence matrix $\boldsymbol{A}$ for the graph in Figure 9.

$$\boldsymbol{A} = \begin{bmatrix} \quad \end{bmatrix}$$

b) Determine the number of paths of length 3 from P2 to P5. Explicitly write out the paths.

c) Determine the number of paths of length 3 or less from P2 to P5.

d) Use routine **igraph** to view a different configuration for the graph in Figure 9. Carefully inspect that there is a one-to-one correspondence between the nodes and the edges in Figure 9 and that displayed by **igraph**. You can use this as a check that your incidence matrix is correct. Sketch the two graphs below and indicate the correspondence by drawing arrows between nodes and marking like edges.

2. A telecommunications network [1] connects any pair of users by means of communication links. In Figure 10 subscribers are denoted S1 through S6; local exchanges are denoted L1 through L3; group switching centers are denoted G1 through G3.

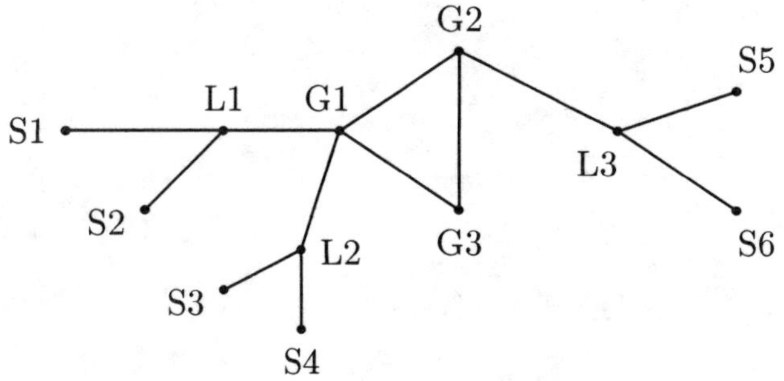

Figure 10

The incidence matrix $\boldsymbol{A}$ for this network is constructed in the standard way: the (i,j)-entry is 1 if there is a direct communication connection between node i and node j, otherwise it is zero. Note that there is no direct connection between subscribers, none between local exchanges, and none between subscribers and group switching centers. Hence the blocks of zeros in the incidence matrix. Construct the remainder of the incidence matrix by filling in its entries and then enter it into MATLAB . (Matrix $\boldsymbol{A}$ is 12×12.)

[1] The network depicted here is commonly called a network 'in the large' because within each switching center and local exchange is another internal network which is much more complex.

$\boldsymbol{A} =$

	$S1$	$S2$	$S3$	$S4$	$S5$	$S6$	$L1$	$L2$	$L3$	$G1$	$G2$	$G3$
$S1$	0	0	0	0	0	0				0	0	0
$S2$	0	0	0	0	0	0				0	0	0
$S3$	0	0	0	0	0	0				0	0	0
$S4$	0	0	0	0	0	0				0	0	0
$S5$	0	0	0	0	0	0				0	0	0
$S6$	0	0	0	0	0	0				0	0	0
$L1$							0	0	0			
$L2$							0	0	0			
$L3$							0	0	0			
$G1$	0	0	0	0	0	0						
$G2$	0	0	0	0	0	0						
$G3$	0	0	0	0	0	0						

a) Without computing $\boldsymbol{A}^2$, describe the set of subscribers that are connected[2] by a path of length 2.

b) Verify that there are no paths of length 3 between any two subscribers. Explain why.

c) What is the longest path required so that any subscriber Si can connect with any other subscriber Sj?

[2]The entries of the powers of the incidence matrix count the number of paths between nodes in the network. But not all of those paths are physically meaningful for communication. For example, the path of length 4 from S1 to S2 that is given by S1 to L1 to G1 to L1 to S2 traverses the link from L1 to G1 twice and is called a 're-entrant path'. There are important questions in communication networks about the number of distinct paths and the number of non-re-entrant path. This area is known as connectivity in graph theory.

d) Which group exchange could be removed and not affect communication services? Explain your choice.

e) Discuss the 'weak points' of this network. That is, those nodes which if rendered inoperable prevent completion of subscriber calls.

3. A **digraph** is a graph whose edges are directed. That is, edges may be one-way paths between nodes. At this time we will also permit nodes to be connected by more than one edge. See Figures 11 and 12.

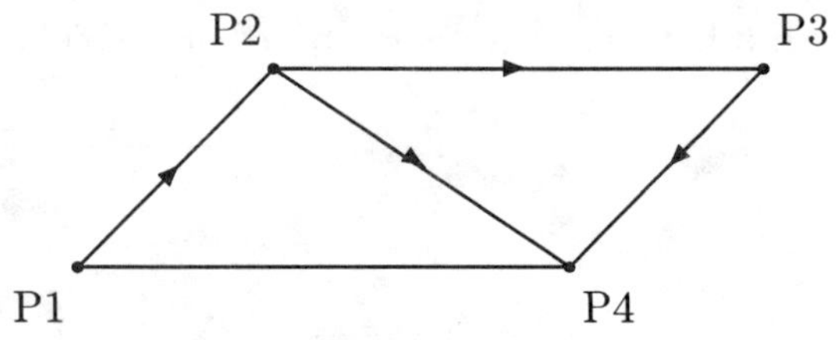

Figure 11

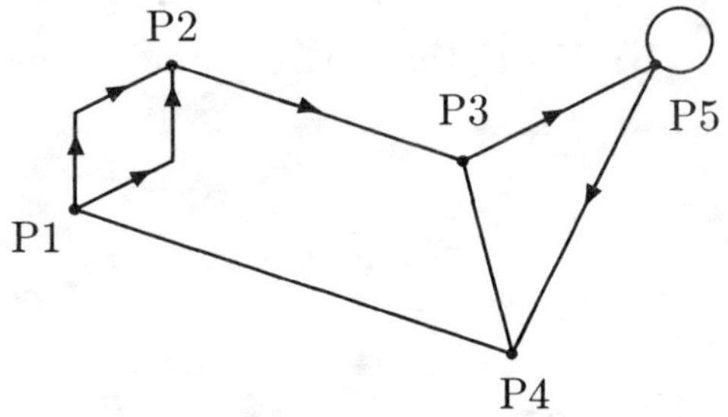

Figure 12

In Figure 11 the edge connecting P1 and P4 is two way since no direction arrow appears on it. In Figure 12 there are two edges connecting P1 to P2 and there is a 'loop' connecting P5 to P5.

With each digraph we associate an incidence matrix $\boldsymbol{A}$ where

$$a_{ij} = \text{the number of edges connecting Pi to Pj}$$

The entries are not restricted to be just 0 or 1 and a_{jj} need not be zero since loops are permitted. In fact the (5,5)-entry of the incidence matrix for Figure 12 is 2 since the loop is not directed.

a) Construct the incidence matrix $\boldsymbol{A}$ for the graph in Figure 11.

$$\boldsymbol{A} = \begin{bmatrix} \quad \\ \quad \\ \quad \end{bmatrix}$$

b) Construct the incidence matrix $\boldsymbol{B}$ for the graph in Figure 12.

$$\boldsymbol{B} = \begin{bmatrix} \quad \\ \quad \\ \quad \end{bmatrix}$$

c) Determine the length of the longest path required between any two nodes for the graph in Figure 11.

d) Determine the length of the longest path required between any two nodes for the graph in Figure 12.

4. Let $\boldsymbol{A} = \begin{bmatrix} 0 & 2 & 0 & 1 \\ 0 & 1 & 1 & 1 \\ 1 & 0 & 0 & 1 \\ 1 & 0 & 1 & 0 \end{bmatrix}$ be the incidence matrix of a digraph.

a) Construct the digraph. (Note: routine **igraph** will not draw digraphs.) Display it below.

b) Find the number of paths of length 2 from P1 to P4.

c) Determine the length of the longest path required between any two nodes.

5. The VAN service provides one-way transportation or two-way transportation between the 'retail outlets' labeled P1 to P6 in Figure 13.

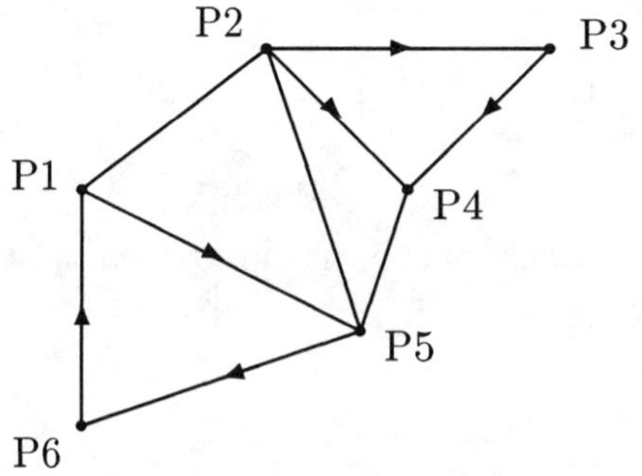

Figure 13

a) Construct the incidence matrix $\boldsymbol{A}$ associated with the graph in Figure 13.

$$A = \begin{bmatrix} \\ \\ \\ \\ \end{bmatrix}$$

b) Determine the number of paths of length 3 from P2 to P5. List them explicitly.

c) What is the length of the shortest path from P3 to P2? ______________.

d) Determine a path from P3 to P2 that goes through P1.

e) What new nondirect route could be added to improve the service from P3 to P1? Explain your choice.

6. A **dominance-directed** graph is a digraph with no loops and with at most one path of length 1 either from Pi to Pj or from Pj to Pi, but not both, for $i \neq j$.

a) What are the diagonal elements of an incidence matrix associated with a dominance directed graph?

____

b) Can the incidence matrix of a dominance-directed graph have two-way paths? Explain.

---

c) The matrix $A = \begin{bmatrix} 0 & 1 & 0 & 1 \\ 0 & 0 & 1 & 1 \\ 1 & 0 & 0 & 0 \\ 0 & 0 & 1 & 0 \end{bmatrix}$ is an incidence matrix of a dominance-directed graph.
Construct the graph below.

7. A simple situation that can be modeled by a dominance-directed graph is a round-robin tournament in which each team plays every other team exactly once. There is a path from Pi to Pj if team i beats (dominates) team j. Let the graph in Figure 14 represent a round-robin tournament among four softball teams.

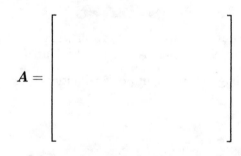

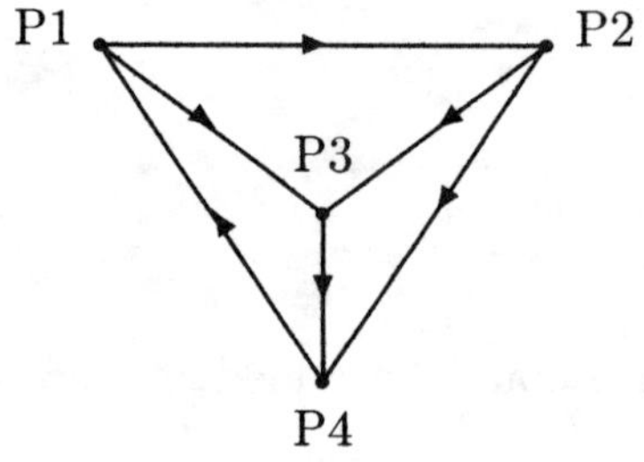

Figure 14

a) Construct the incidence matrix A for the graph in Figure 14.

$$A = \begin{bmatrix} & & & \\ & & & \\ & & & \\ & & & \end{bmatrix}$$

b) A node Pj is called **powerful** in a dominance-directed graph provided there is a path of length 1 or length 2 from Pj to every other node. This implies that either team j beats every other team or beats a team that has defeated a team that team j did not defeat. Which, if any, are the powerful teams in the graph in Figure 14?

c) The **power** of a node Pj is the number of paths of length 1 or 2 from Pj to all other nodes. Compute the power of each node for the graph in Figure 14.

<< NOTES; COMMENTS; IDEAS >>

<< NOTES; COMMENTS; IDEAS >>

Secret Codes

Description: An application of matrix concepts to encoding and decoding messages. Messages are handled alphabetically using modulo arithmetic in MATLAB .

Prerequisites: matrix multiplication; matrix inverse.

PROJECT 2

CYEY-JPD is a message in a kind of foreign language – not a typical language, but a code instead. The goal is to decode this message.

The word 'cryptography' refers to the encoding of messages and the deciphering of codes. One of the easiest ways to encode a message is to identify each letter with the number of its position in the alphabet, shown here with the number below each letter and with the number 0 for a dash.

-	A	B	C	D	E	F	G	H	I	J	K	L	
0	1	2	3	4	5	6	7	8	9	10	11	12	
M	N	O	P	Q	R	S	T	U	V	W	X	Y	Z
13	14	15	16	17	18	19	20	21	22	23	24	25	26

In this scheme the message ATTACK is written 1 20 20 1 3 11 .

Such a method of encoding is too easy to decode. On the other hand, matrices supply the perfect structure for holding numbers in place, and matrix multiplication provides an easy method for encoding messages. We will show how matrices and their inverses are used in cryptography.

Encoding a Message

The first example will use columns with two entries. The method for encoding a message begins with an encoding matrix $\boldsymbol{E}$. Here let

$$\boldsymbol{E} = \begin{bmatrix} 5 & 8 \\ 8 & 13 \end{bmatrix}$$

Enter matrix $\boldsymbol{E}$ into MATLAB . The matrix corresponding to the message ATTACK is $\begin{bmatrix} 1 & 20 & 20 & 1 & 3 & 11 \end{bmatrix}$. We will call this the **input message**, and denote it by the name **inmess**. In MATLAB type

$$\textbf{inmess} = \begin{bmatrix} 1 & 20 & 20 & 1 & 3 & 11 \end{bmatrix}$$

Convert this 1×6 matrix to a 2×3 matrix $\boldsymbol{M}$, called the **message matrix**, by typing the command

M = reshape(inmess,2,3)

Verify that the message matrix for ATTACK is

$$\boldsymbol{M} = \begin{bmatrix} 1 & 20 & 3 \\ 20 & 1 & 11 \end{bmatrix}$$

The method for encoding a message is to form the product of the encoding matrix $\boldsymbol{E}$ and the message matrix $\boldsymbol{M}$, so type

$$\boldsymbol{A}{=}\boldsymbol{E}{*}\boldsymbol{M}$$

Verify that

$$\boldsymbol{A} = \begin{bmatrix} 165 & 108 & 103 \\ 268 & 173 & 167 \end{bmatrix}$$

The entries in matrix $\boldsymbol{A}$ cause a slight problem because they are not integers between 0 and 26. The method for determining the corresponding letters is to divide 27 into each entry and form the remainder.[1] For instance, for the first entry 165 the remainder when divided by 27 is 3 so 165 corresponds to the letter C. The command in MATLAB that performs this arithmetical drudgery for you is **modn**. Type

C = modn(A,27)

This is called the **code-matrix**. Verify that

$$\boldsymbol{C} = \begin{bmatrix} 3 & 0 & 22 \\ 25 & 11 & 5 \end{bmatrix}$$

The **reshape** command converts C to the **output message**, which we name **outmess**. Type

outmess = reshape(C,1,6)

The resulting matrix is $\begin{bmatrix} 3 & 25 & 0 & 11 & 22 & 5 \end{bmatrix}$. By interpreting each number as a letter we see that the code for ATTACK is CY-KVE. Notice that two different letters represent A and two different symbols represent T. Thus this method negates a statistical analysis that relies on the frequency of occurrence of certain letters.

[1] Technically this is a representation using modulo 27 arithmetic. The number 27 accounts for the 26 letters and a dash

Method for encoding a message with m characters.

1. Enter an n × n encoding matrix $\boldsymbol{E}$.
2. Let **inmess** be the row matrix of numbers corresponding to the m letters of the message. Insert dashes for spaces. If needed, add enough dashes to the message so that its new length k is evenly divisible by n.
3. Form the message matrix **M = reshape(inmess,n,k/n)** using numerical values for k and n.
4. Compute **A = E*M**.
5. Convert matrix $\boldsymbol{A}$ to the code matrix using command **C = modn(A,27)**.
6. Let **outmess = reshape(C,1,k)** and interpret its entries as letters.

Decoding a Message

Now we are ready to decipher (literally meaning 'from the numbers') the code CYEY-JPD that began this unit. First, enter the output message by typing

outmess = [3 25 5 25 0 10 16 4]

The code matrix $\boldsymbol{C}$ must have two rows because $\boldsymbol{E}$ is a 2 × 2 matrix. Since **outmess** has 8 entries, $\boldsymbol{C}$ must have 4 columns because 8/2 = 4. Form $\boldsymbol{C}$ by typing

C = reshape(outmess,2,4)

Verify that

$$\mathbf{C} = \begin{bmatrix} 3 & 5 & 0 & 16 \\ 25 & 25 & 10 & 4 \end{bmatrix}$$

The goal is to find the message matrix $\boldsymbol{M}$ such that $\boldsymbol{EM} = \boldsymbol{C}$. The solution is to form $\boldsymbol{B} = \boldsymbol{E}^{-1}\boldsymbol{C}$ and then convert $\boldsymbol{B}$ to a matrix $\boldsymbol{M}$ whose entries are integers between 0 and 26. Type

B = invert(E)*C

Verify that

$$\boldsymbol{B} = \begin{bmatrix} -161 & -135 & -80 & 176 \\ 101 & 85 & 50 & -108 \end{bmatrix}$$

What letter corresponds to -161? To find the letter you move backwards through the alphabet (including the dash) 5 times to account for 135 letters. The letter that corresponds to -161 is found by moving backwards 26 more letters. Therefore the letter is A. Once again, MATLAB can perform this drudgery for you. Type

M = modn(B,27)

Verify that the message matrix is

$$M = \begin{bmatrix} 1 & 0 & 1 & 14 \\ 20 & 4 & 23 & 0 \end{bmatrix}$$

The command

inmess = reshape(M,1,8)

produces a string of numbers that converts to the message AT-DAWN-. Notice that the first letter, A, corresponds to the number -161.

Note the double use for the dash in the preceding message. The first dash is used as a word separator; the second to complete the final column of M in the code, since otherwise the last column of M would have had only one entry.

Method for decoding a message with m characters.

1. Enter an n $\times$ n encoding matrix E.
2. Let **outmess** be the row matrix of numbers corresponding to the m letters of the code. Insert dashes for spaces. If needed, add enough dashes so that its new length k is evenly divisible by n.
3. Form the code matrix using **C = reshape(outmess,n,k/n)** using numerical values for k and n.
4. Compute **B = invert(E)*C**.
5. Convert **B** to the message matrix using **M = modn(B,27)**.
6. Let **inmess = reshape(M,1,k)** and interpret its entries as letters.

Exercises

1. Encode LINEAR ALGEBRA using the encoding matrix $\boldsymbol{E} = \begin{bmatrix} 5 & 8 \\ 8 & 13 \end{bmatrix}$. Put a dash between the two words. Record the letters of the coded message below the original letters.

L I N E A R - A L G E B R A

____

2. Decode YPWBUMBO-ONQ using the same encoding matrix $\boldsymbol{E}$ in Exercise 1. Record your answer below.

____

3. Decode LKSSSOLFGWFXBSBX using the encoding matrix $\boldsymbol{E} = \begin{bmatrix} 1 & 1 \\ 5 & 6 \end{bmatrix}$. Record your answer below.

____

The examples so far have made use of a 2×2 encoding matrix $\boldsymbol{E}$. If $\boldsymbol{E}$ is a 3×3 matrix then the columns of the message matrix $\boldsymbol{M}$ and the code matrix $\boldsymbol{C}$ must have 3 entries. As before, use a dash to separate words and to fill out unused entries in $\boldsymbol{M}$ or $\boldsymbol{C}$.

In Exercises 4 and 5 use encoding matrix $\boldsymbol{E} = \begin{bmatrix} 5 & 4 & 2 \\ 8 & 8 & 9 \\ 4 & 3 & 1 \end{bmatrix}$.

4. Encode LINEAR ALGEBRA and record your answer below.

____

5. Decode PXHKUNAHOEFRMQU and record your answer below.

____

6. Use the encoding matrix $\boldsymbol{E} = \begin{bmatrix} 1 & 0 & 0 & 1 \\ 0 & 1 & 1 & 0 \\ 0 & 1 & 0 & 0 \\ 1 & 0 & 0 & 0 \end{bmatrix}$, to decode

LQYSTMMEWE-FCUAUOFNO. Record your answer below.

____

7. Decode the message FRRBVQCYZMTAJAPB by finding the matrix $\boldsymbol{E}$ which was used to encode the message. It is known that $\boldsymbol{E}$ satisfies the following properties:

1. $\boldsymbol{E}$ is 2×2.
2. Each entry of $\boldsymbol{E}$ is 1 or 2.
3. If $\boldsymbol{E} = \begin{bmatrix} a & b \\ c & d \end{bmatrix}$, then $|ad - bc| = 1$.

Warning: there can be more than 1 matrix satisfying the conditions above. You must find the one that meaningfully decodes your message.

Record the encoding matrix $\boldsymbol{E}$. ____________________

Record the decoded message here. ______________________________

8. A very simple way to encode a message is to reverse successive pairs of the letters. For example, MATRIX is encoded as AMRTXI. What 2×2 matrix $\boldsymbol{E}$ encodes a message in this way?

____

9. A new 3×3 code matrix $\boldsymbol{E} = \begin{bmatrix} 2 & -3 & -5 \\ 0 & 1 & -1 \\ 2 & -1 & -7 \end{bmatrix}$ was received by the station code expert.

As she was using $\boldsymbol{E}$ to encode a message she noticed that

$$\boldsymbol{E} * \begin{bmatrix} 4 \\ 1 \\ 1 \end{bmatrix} = \begin{bmatrix} 0 \\ 0 \\ 0 \end{bmatrix} \qquad \boldsymbol{E} * \begin{bmatrix} 20 \\ 5 \\ 5 \end{bmatrix} = \begin{bmatrix} 0 \\ 0 \\ 0 \end{bmatrix}$$

Write a short paragraph below to explain why this is an unacceptable code matrix.

<< NOTES; COMMENTS; IDEAS >>

Least Squares Models

Description: The least squares line model is developed both geometrically and algebraically. Routine **lsqgame** uses MATLAB 's graphics interface to provide interactive experimentation in estimating a 'line of best fit'. Routine **lsqline** provides a tool for experimentation with data sets and exploring line models. Examples based on real data from Olympic events are used. The notion of a pseudo inverse provides a unifying theme for the development.

Prerequisites: matrix multiplication, transposes, and inverses; linear systems; linear independence; dot product; orthogonal vectors; column space.

PROJECT 3

An important use for linear systems of equations is in the construction of mathematical models for a set a data. The mathematical model is an expression that provides a relationship between the coordinates of the data. The simplest instance is when the data consist of ordered pairs (x_i, y_i) and there is a 'linear' relationship between x_i and y_i. That is, there is some line $y = mx + b$ which either goes through each pair (x_i, y_i) or comes close to all the pairs. If each data pair lies on the same line then values for m and b can be determined from the solution of a nonsingular linear system. If any data pair does not lie on the same line then an inconsistent system arises in the model building process. We show how to develop a related nonsingular system whose solution provides a 'linear' model that lies *close* to all the data. The model we develop is called the **least squares line model**.

Section 1 provides a geometric setting for least squares line model using MATLAB 's graphical user interface. The routine **lsqgame** actively engages the student in experiments to build linear models. Students can compete against one another and the routine has an option to select data sets using a mouse. This section can be used independently from the other sections and requires very few linear algebra topics.

Section 2 develops least squares line models using matrix algebra and the notion of projections. Projection properties are introduced informally and we do not assume that Lab 10 has been covered. The pseudo inverse is introduced and provides a unifying theme for other types of models developed later. We use MATLAB 's graphics in routine **lsqline** to provide visualization of the line models.

This Project contains both geometric and algebraic approaches to least squares model building. There is also a wide variety of exercises using data of various types. We have used these materials with lab groups and as individual lab projects for students. There are plenty of ideas and materials to provide a flexible approach to the study of and experimentation with least squares.

Section 1.

Geometric Experimentation with Linear Models

A primary focus has been linear systems of equations. Consistent linear systems provide an important modeling tool for a variety of applications. The solution set of a linear system provides us with values for the unknowns that *exactly* satisfy each equation. For instance, suppose that we have collected data from a process by supplying values for an input parameter and recording values of an output, or response. We use the data in Table 1.

x – input	y – output
2.1	4.9200
3.2	8.4400
−2.7	−10.4400
4.2	11.6400
0.53	−0.1040

Table1.

If we suspect that there is a linear relationship between x and y this implies that there exist values m and b so that

$$y = mx + b$$

for each input-output pair (x, y) in Table 1. Substitute the data pairs from Table 1 into the preceding equation model to develop a system of equations to solve for m and b. We get

$$\begin{array}{rcrcr} 2.1 * m & + & b & = & 4.9200 \\ 3.2 * m & + & b & = & 8.4400 \\ -2.7 * m & + & b & = & -10.4400 \\ 4.2 * m & + & b & = & 11.6400 \\ 0.53 * m & + & b & = & -0.1040 \end{array}$$

In matrix form we have

$$\begin{bmatrix} 2.1 & 1 \\ 3.2 & 1 \\ -2.7 & 1 \\ 4.2 & 1 \\ 0.53 & 1 \end{bmatrix} \begin{bmatrix} m \\ b \end{bmatrix} = \begin{bmatrix} 4.9200 \\ 8.4400 \\ -10.4400 \\ 11.6400 \\ -0.1040 \end{bmatrix}$$

Forming the augmented matrix of this system and computing the rref we get

$$\textbf{rref}\left(\begin{bmatrix} 2.1 & 1 & 4.9200 \\ 3.2 & 1 & 8.4400 \\ -2.7 & 1 & -10.4400 \\ 4.2 & 1 & 11.6400 \\ 0.53 & 1 & -0.1040 \end{bmatrix}\right) = \begin{bmatrix} 1 & 0 & 3.2 \\ 0 & 1 & -1.8 \\ 0 & 0 & 0 \\ 0 & 0 & 0 \\ 0 & 0 & 0 \end{bmatrix}$$

which implies that the data is 'exactly' modeled by $y = 3.2x - 1.8$. We say that this line *'fits the data'*. Geometrically we interpret this to say that the five data pairs are collinear.

Suppose that our instruments to measure the response y had a small malfunction and we could only obtain output information as whole number values. Table 2 contains this data where we represent the response by the unknown z.

x − input	z − output
2.1	4
3.2	8
−2.7	−10
4.2	11
0.53	0

Table 2.

Following the model building procedure above we get linear system

$$\begin{bmatrix} 2.1 & 1 \\ 3.2 & 1 \\ -2.7 & 1 \\ 4.2 & 1 \\ 0.53 & 1 \end{bmatrix} \begin{bmatrix} m \\ b \end{bmatrix} = \begin{bmatrix} 4 \\ 8 \\ -10 \\ 11 \\ 0 \end{bmatrix}$$

Forming the augmented matrix of this system and computing the rref we get

$$\mathbf{rref}\left(\begin{bmatrix} 2.1 & 1 & 4 \\ 3.2 & 1 & 8 \\ -2.7 & 1 & -10 \\ 4.2 & 1 & 11 \\ 0.53 & 1 & 0 \end{bmatrix} \right) = \begin{bmatrix} 1 & 0 & 0 \\ 0 & 1 & 0 \\ 0 & 0 & 1 \\ 0 & 0 & 0 \\ 0 & 0 & 0 \end{bmatrix}$$

It follows that this system is inconsistent; that is, there is no line that goes through all the data pairs (x, z) from Table 2. However, at least intuitively, we feel there is some line that should come 'close' to all these points. To determine the closest line to the data in Table 2 we must somehow use the inconsistent system developed above. Since the system is inconsistent we can not expect to satisfy each of the five equations exactly, hence we need to change our point of view from 'equality' to 'approximately equal'.

The key to finding the line closest to the data set is to proceed both geometrically and algebraically as we illustrate next. In Figures 1 and 2 the same data set is displayed together with a line that is close to all the data. (It is an optical illusion that the data sets appear different.) By observation we see that no line can go through all the points. But which of the lines in Figures 1 and 2 is 'closer' to all the points?

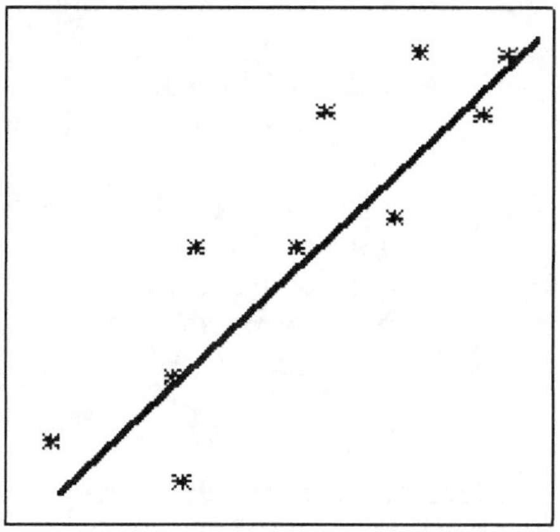

Figure 1.

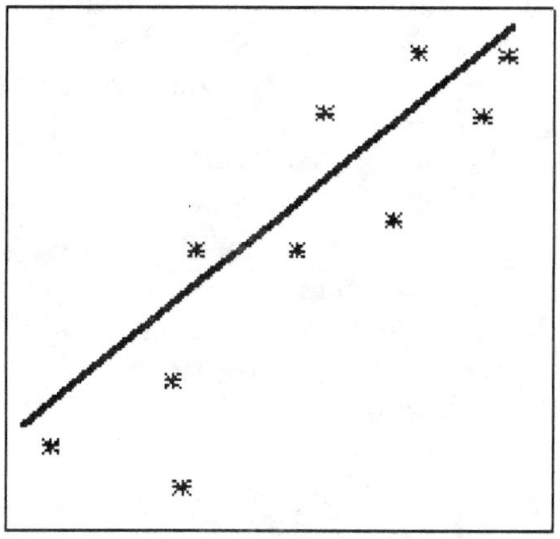

Figure 2.

To measure 'closeness' we compute the vertical distance from each point to the line chosen to be the model. Then we take the sum of the squares of these *deviations* as our measure of closeness. (Section 2 develops this procedure in detail.) The line for which the sum of the squares of the (vertical) deviations is a minimum is called **the Least Squares Line** for the data set. Figure 3 shows the least squares line and the deviations for the data set used in Figures 1 and 2.

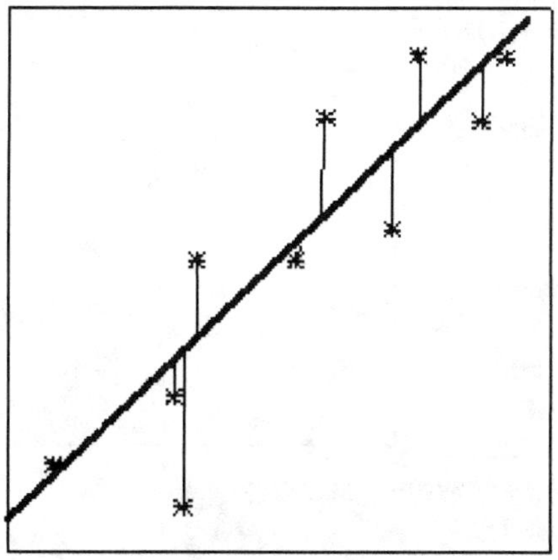

Figure 3.

The equation of the least squares line shown in Figure 3 is obtained from the solution of a linear system that is *'related'* to the inconsistent system that would result if we substituted each data pair into equation $y = mx + b$. The solution procedure to determine m and b is developed in Section 2.

Here we proceed geometrically to provide a feel for the least squares process. The routine **lsqgame** lets you use a mouse to select a 'line model' for a data set. In this routine you click on the graphics display of the data set to select two points for a line. The routine shows the vertical deviations, computes the sum of the squares of the deviations, and determines the equation of the line through the points you selected. A second guess for the least squares line can be made in order to try to reduce the sum of the squares of the deviations so you produce a line model that is closer to the data set. One way to sharpen your estimation skills is to compete with a classmate to see who gets the 'better fit' to the data. (Hence the designation **lsqgame**. For instance, flip a coin to see who selects the data set or have a third neutral party select the data). The data can be selected by the mouse or through several other options; type **help lsqgame** for more information. Next decide who makes the first estimate of the least squares line. The better estimate is the one with the smaller sum of the squares of the deviations. There is an option to show the least squares line. To initiate the game type **lsqgame**, follow the on-screen directions, and click the appropriate buttons. Figure 4 shows a display from **lsqgame** with a guess at the least squares line.

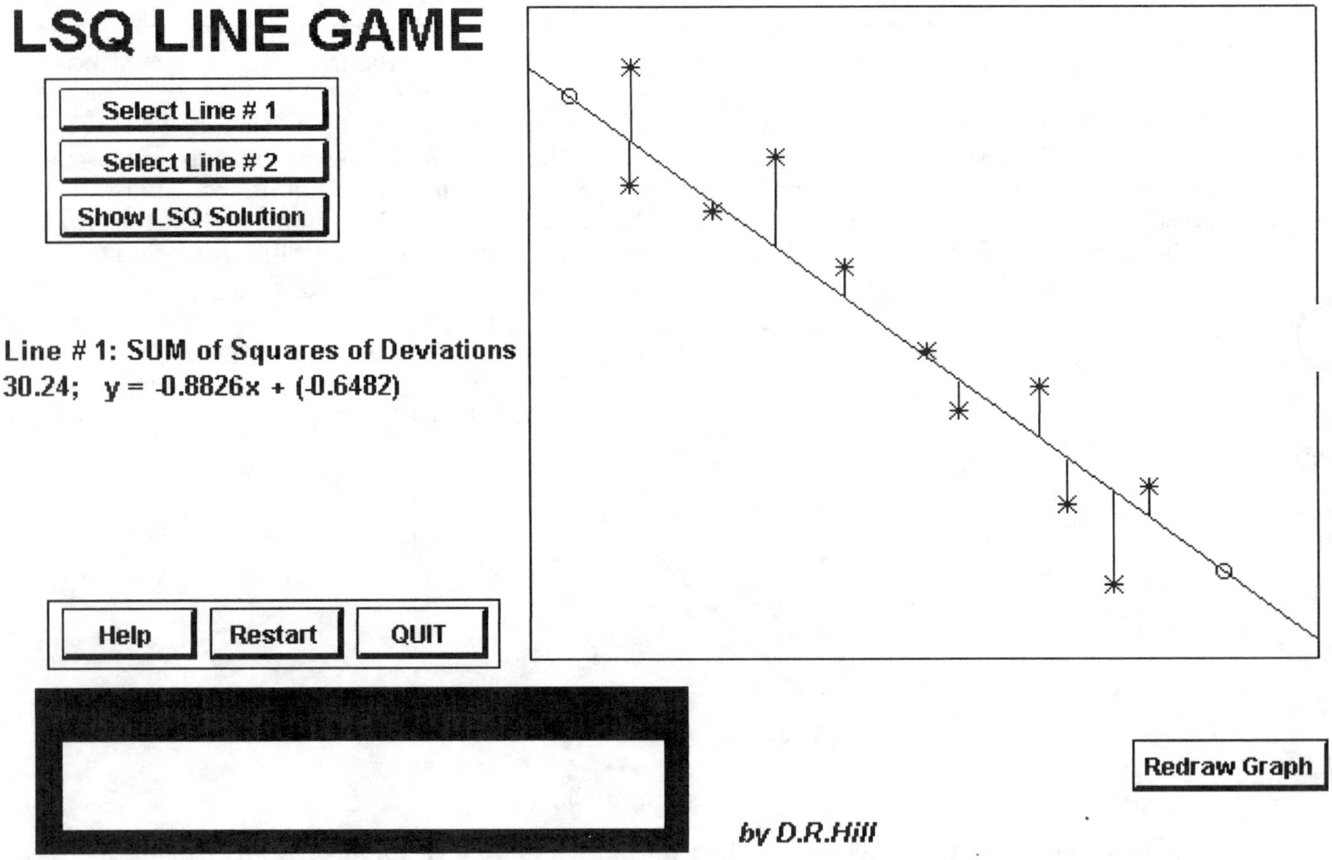

Figure 4.

Example 1. To illustrate routine **lsqgame** we use the data from the men's Olympic pole vault event. For convenience we have stored this data in a file which can be loaded into MATLAB . Enter the following command to load and display the data.

load pvault
data = [x y]

The year is listed in the first column and the winning height (in inches) in the second column. To develop a line to fit this data type **lsqgame**. After you select the number of players, 1 or 2, choose option 2 at the Data Entry Menu. At the entry prompt type **data** and press ENTER. Then follow the screen directions to generate your line to fit this data.

Record the equation of the least squares line approximation to the pole vault data on the line below.

____

Click on the QUIT button to return to the MATLAB prompt. Then use this equation to predict the winning pole vault height in inches for the 2012 Olympics. Show your work in the space below.

The pole vault data are shown in Figure 5. In the space below write a short description of how well you feel this data can be approximated by a single line. Be specific.

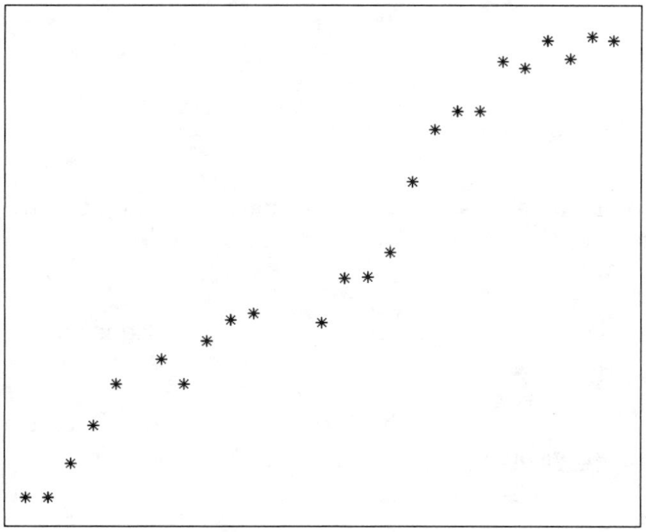

Figure 5.

Example 2. Execute **lsqgame** choosing 1 or 2 players as appropriate. For DATA ENTRY choose Option 1 to use the mouse to select a data set. Follow the on-screen directions. Note that to quit selecting points press *q* on the keyboard. Use the mouse to select a data set which resembles that displayed in Figure 6. Choose as the final piece of data a point similar to the one designated 'last' in Figure 6.

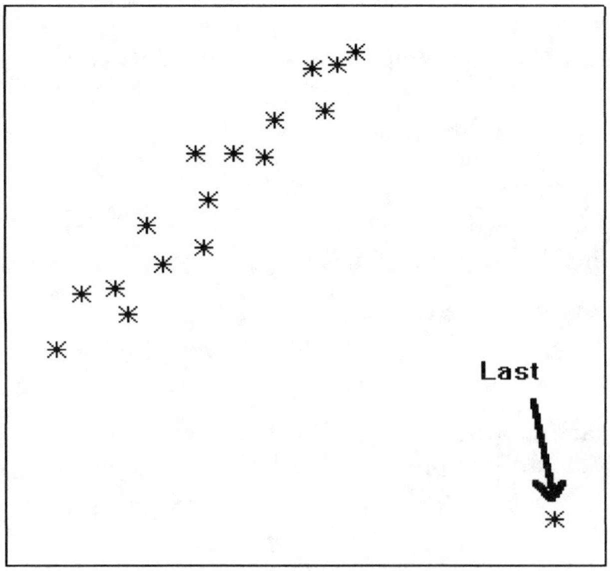

Figure 6.

Construct your estimate of the least squares line to data set you generated. Record the sum of squares of the deviations and the equation of your line below.

Sum of squares of deviations: __________ Equation: __________

Have **lsqgame** compute the least squares line to your data set. Record the sum of squares of the deviations and its equation below.

Sum of squares of deviations: __________ Equation: __________

Sketch your data set, your estimate of the least squares line, and the least squares line in the box of Figure 7.

Figure 7.

Click on the QUIT button. At the command screen prompt type **dmat**. You will see a table of x and y values of the data set you created with the mouse. The 'last' data point is so different from the others, the question that arises is, *How does the least squares model change if we omit that'last' point?* To aid in the investigation of this question use the following commands to drop the 'last' data point, and then recompute the models. (In the following commands m is the number of data point you selected, and matrix **newdata** will contain the original set except for the 'last' point.)

[m,n] = size(dmat)
newdata = dmat(1:m-1,:)

Start **lsqgame**, select the appropriate number of players, and choose option 2 for DATA ENTRY. At the data input prompt type **newdata**. Record the sum of squares of the deviations and the equation of your line below.

Sum of squares of deviations: ____________ Equation: ____________

Have **lsqgame** compute the least squares line to your data set. Record the sum of squares of the deviations and its equation below.

Sum of squares of deviations: ____________ Equation: ____________

In the space below describe the effect of the 'last' point on the least squares line models generated above. (The point we designated as 'last' is called an *outlier*.)

Click on QUIT to return to the MATLAB prompt.

Exercises 1.

1. The cost of health care insurance in the United States for the years 1994 - 2001 in billions of dollars is given in the following table. Create an 8×2 matrix **hdata** in MATLAB with the year in column 1 and the expenditure in column 2. Use **lsqgame** to estimate a least squares fit for the data. Choose option 2 for input so that you can just type the name **hdata** to create the data set.

year	1994	1995	1996	1997	1998	1999	2000	2001
Health Ins. Cost	55.8	58.0	56.6	59.3	63.6	65.7	70.6	75.0

Is this data well approximated by a straight line? Explain.

Use the least squares line equation to predict the expenditure in 2010. What assumption is implicit in this calculation?

2. The Consumer Price Index (CPI) is a measure of the average change in prices over time of selected goods and services. To establish a base for comparison the annual average of monthly prices in the years 1982-84 have been used to generate data in the following table of consumer price changes for tobacco products. The figures represent the percentage of change from the base.

year	1970	1975	1980	1985	1990	1995	1999	2000	2001
Tobacco CPI	43.1	54.7	72.0	116.7	181.5	225.7	355.8	374.9	425.2

Is this data well approximated by a straight line? Explain.

Use the least squares line equation to predict the tobacco CPI in 2010. What assumption is implicit in this calculation?

3. Each year the U.S. government spends lots of your money on the national debt. The following table shows the percent of government expenditures that goes towards paying interest on the national debt for the years 1991-2002.

year	1991	1993	1995	1997	1999	2001	2002
% of govt. expend.	21.6	20.8	22.0	22.2	20.7	19.3	16.4

Is this data well approximated by a straight line? Explain.

Use the least squares line equation to predict the percent of federal outlays applied to the national debt in 2010. What assumption is implicit in this calculation?

4. The United States census is taken every ten years. The following table shows the census figures from 1940 through 2000 in millions (rounded).

year	1940	1950	1960	1970	1980	1990	2000
U.S.population	132	151	179	203	227	249	281

Is this data well approximated by a straight line? Explain.

Use the least squares line equation to estimate the U.S. population in 2010 and in 2020.

Use the least squares line equation to estimate the year when the U.S. population will reach 320 million.

5. The length in inches of the winning jumps in the men's long jump for Olympic competition is loaded using command **longjump**. Use **help longjump** for details. Use **lsqgame** to estimate a least squares fit for this data. Discuss the possibility of an outlier in this data set. Explain.

Section 2.

The Linear Model

The women's 100-meter dash was won in flamboyant style by Florence 'Flo-Jo' Griffith-Joyner at the 1988 Olympics in Seoul, Korea. Her winning time was 10.54 seconds. We will show how to predict the winning time for this running event in future Olympics based on the winning times in all Olympics since the sprint was first contested.

The women's 100-meter dash was first contested in the Olympics in 1928. It was won by Elizabeth Robinson of the U.S. in a time of 12.2 seconds. The event has been contested at every Olympic Games since then. Table 1 lists the gold medal winners and their times [1] (in seconds.)

Year	Gold Medal Winner	Time
1928	Elizabeth Robinson, U.S.	12.2
1932	Stella Walsh, Poland	11.9
1936	Helen Stephens, U.S.	11.5
1948	Francina Blankers-Koen, Netherlands	11.9
1952	Marjorie Jackson, Australia	11.5
1956	Betty Cuthbert, Australia	11.5
1960	Wilma Rudolf, U.S.	11.0
1964	Wyomia Tyus, U.S.	11.4
1968	Wyomia Tyus, U.S.	11.0
1972	Renate Stecher, E. Germany	11.07
1976	Annegret Richter, W. Germany	11.08
1980	Lyudmila Kondratyeva, USSR	11.6
1984	Evelyn Ashford, U.S.	10.97
1988	Florence Griffith-Joyner, U.S.	10.54
1992	Gail Devers, U.S.	10.82
1996	Gail Devers, U.S.	10.94

TABLE 1.

Table 1 shows that the winning times have been lowered regularly but not steadily (monotonically). (One might argue that the slow time in 1980 was influenced more by politics than sports.) To view the data graphically enter the years as a column matrix in MATLAB via the command

$$\mathbf{x} = [1928\ 1932\ 1936\ 1948{:}4{:}1996]'$$

[1]The years 1940 and 1944 do not appear because no games were held during World War II. All Olympic data were taken from **The World Almanac and Book of Facts 2003**, World Almanac Books, AWRC Media Company 512 Seventh Avenue New York, NY 10018.

Enter the winning times from Table 1 into MATLAB as a column matrix **y**. (There is no painless way to do this.) Then type command

lsqline

From the input options choose 2, and follow the on-screen directions. When the graphics screen appears you will see the data plotted in (x, y)-pairs denoted by $*$ and a line that, in some sense, comes close to the data pairs. Click on the Show Data Table button; check that you entered the data correctly. Follow the screen directions to return to the graphics screen. We will explain the meaning of the line displayed and derive a method for computing its equation.

Ideally we seek a linear relationship between the 16 by 1 matrices **x** (in years) and **y** (in seconds). That is, we seek scalars c_1 and c_2 such that

$$c_1\mathbf{x} + c_2 = \mathbf{y}$$

The equation below the graph reveals that $c_1 = -0.017583$ and $c_2 = 45.8499$. For now click on the QUIT button to exit the routine **lsqline**.

The scalars c_1 and c_2 are derived by dealing with the system of 16 equations in two unknowns displayed next:

$$\begin{aligned}
1928c_1 + c_2 &= 12.2\\
1932c_1 + c_2 &= 11.9\\
1936c_1 + c_2 &= 11.5\\
1948c_1 + c_2 &= 11.9\\
1952c_1 + c_2 &= 11.5\\
1956c_1 + c_2 &= 11.5\\
1960c_1 + c_2 &= 11.0\\
1964c_1 + c_2 &= 11.4\\
1968c_1 + c_2 &= 11.0\\
1972c_1 + c_2 &= 11.07\\
1976c_1 + c_2 &= 11.08\\
1980c_1 + c_2 &= 11.6\\
1984c_1 + c_2 &= 10.97\\
1988c_1 + c_2 &= 10.54\\
1992c_1 + c_2 &= 10.82\\
1996c_1 + c_2 &= 10.94
\end{aligned} \tag{3.1}$$

Enter the coefficient matrix $\boldsymbol{A}$ of system (3.1) into MATLAB using the command

A = [x ones(size(x))]

The augmented matrix of the linear system in Equation (3.1) is [**A y**]. Display its rref by using MATLAB command

rref([A y])

Omitting any zero rows, record the rref of the augmented matrix in the space below.

$$\begin{bmatrix} \quad & \quad \\ \quad & \quad \end{bmatrix}$$

Is the system consistent or inconsistent? (circle one)

consistent inconsistent

You should have observed that the system is inconsistent since the last nonzero row of the augmented matrix had the form [0 0 | 1]. Thus **there is no set of values for c_1 and c_2 which satisfies all 16 equations simultaneously.** However, "In spite of their unsolvability, inconsistent equations arise in practice and have to be solved." [2]

We do not solve an unsolvable system, but instead settle for an approximate solution which we judge to be **'best'** in some sense. In this unit we show how to obtain an approximate solution to an inconsistent system that is **'best in the least squares sense.'**

Let $\boldsymbol{u} = \begin{bmatrix} c_1 \\ c_2 \end{bmatrix}$. Then the linear system in Equation (3.1) has the matrix form

$$\boldsymbol{Au} = \boldsymbol{y} \tag{3.2}$$

Since this system is inconsistent, we know that vector $\boldsymbol{y}$ is not in the column space of matrix $\boldsymbol{A}$, which is span{**x, ones(size(x))**}. While the column space of $\boldsymbol{A}$ is a two dimensional subspace of R^{16} and cannot be displayed geometrically, we intuitively use the corresponding situation in R^3 to serve as motivation on how to proceed. The corresponding situation in R^3 is that we have a plane P (a two dimensional subspace) and a vector $\boldsymbol{y}$ not in the plane as depicted in Figure 1. Since $\boldsymbol{y}$ is not in the plane P we cannot express it as a linear combination of vectors in a spanning set (or basis) for P. Hence we try to find the vector $\boldsymbol{z}$ in P that is 'closest' to $\boldsymbol{y}$. We then replace $\boldsymbol{y}$ by $\boldsymbol{z}$ and solve the related system $\boldsymbol{Aw} = \boldsymbol{z}$. The vector $\boldsymbol{w}$ is considered an approximate

[2] Gilbert Strang, **Linear Algebra and Its Applications**, Academic Press, 1976, p. 105; second edition, 1980, p. 111. In **Introduction to Linear Algebra**, Wellesley-Cambridge Press, 1993, Strang writes (p. 179), "But these are real problems and they need an answer."

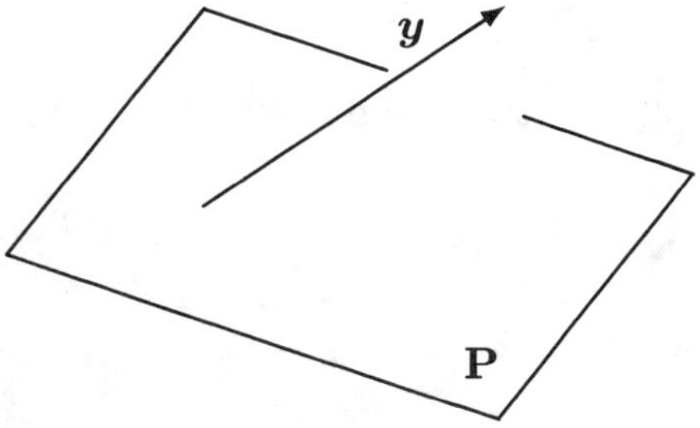

Figure 1

solution to the linear system $\boldsymbol{Au} = \boldsymbol{y}$. From an intuitive geometrical point of view we will find $\boldsymbol{z}$ by dropping a perpendicular from the tip of $\boldsymbol{z}$ to the plane P. See Figures 2 and 3. If we label the perpendicular $\boldsymbol{n}$ as in Figure 3, then the vector in P that is closest to $\boldsymbol{y}$ is $\boldsymbol{z} = \boldsymbol{y} - \boldsymbol{n}$. Since the figures are in R^3, it is natural to expect that we can use algebra and trigonometry to determine vectors $\boldsymbol{n}$ and $\boldsymbol{z}$. A careful analysis shows that this development can be done using matrix algebra and the notion of orthogonal vectors and the development can be immediately generalized to R^{14} or in general R^k .

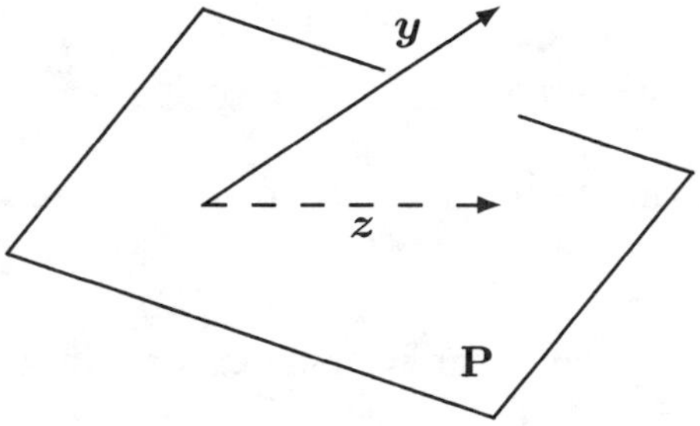

Figure 2

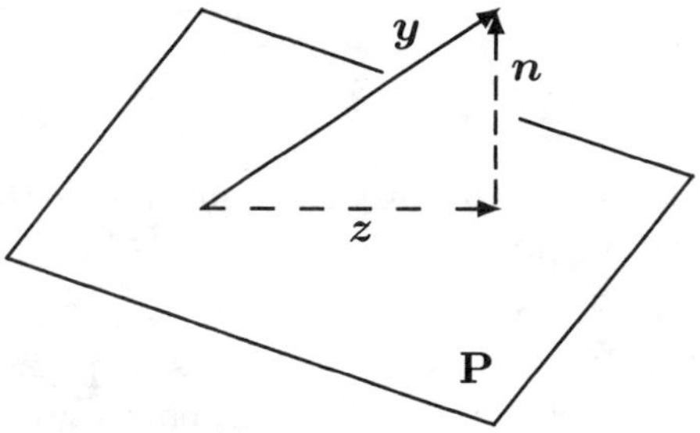

Figure 3

Following is a step-by-step argument of how to convert the geometry displayed in Figure 3 into matrix algebra so that we can compute vector $\boldsymbol{w}$ explicitly from the coefficient matrix $\boldsymbol{A}$ and right-hand side $\boldsymbol{y}$ in Equation (3.2). We let P denote the column space of $\boldsymbol{A}$ for ease in referring to Figure 3.

- $\boldsymbol{n}$ is orthogonal to subspace P. Equivalently, $\boldsymbol{n}$ is orthogonal to every vector in subspace P. (This fact is often called the projection property.)

- $\boldsymbol{n}$ is orthogonal to the spanning set of subspace P.

- $\boldsymbol{n}$ is orthogonal to each column of matrix $\boldsymbol{A}$. (In our situation there are only two columns.)

- In terms of dot products, $\text{col}_1(\boldsymbol{A})\cdot\boldsymbol{n} = 0$ and $\text{col}_2(\boldsymbol{A})\cdot\boldsymbol{n} = 0$.

- In terms of matrix products, $\text{col}_1(\boldsymbol{A})^T{*}\boldsymbol{n} = 0$ and $\text{col}_2(\boldsymbol{A})^T{*}\boldsymbol{n} = 0$.

- Using properties of matrix algebra we combine these two equations into the single matrix equation

$$\begin{bmatrix} \text{col}_1(\boldsymbol{A})^T \\ \text{col}_2(\boldsymbol{A})^T \end{bmatrix} * \boldsymbol{n} = \begin{bmatrix} 0 \\ 0 \end{bmatrix} = \boldsymbol{0}$$

which is the same as the matrix equation $\boldsymbol{A^T} * \boldsymbol{n} = \boldsymbol{0}$.

- From Figure 3 we see that $\boldsymbol{n} = \boldsymbol{y} - \boldsymbol{z}$ and since $\boldsymbol{z}$ is in subspace P it is a linear combination of the columns of matrix $\boldsymbol{A}$. Hence we can express vector $\boldsymbol{z}$ as $\boldsymbol{z} = \boldsymbol{A} * \boldsymbol{w}$, where $\boldsymbol{w}$ is a column of coefficients that express $\boldsymbol{z}$ as a linear combination of the columns of $\boldsymbol{A}$. Thus we have the matrix equation

$$\boldsymbol{A^T} * \boldsymbol{n} = \boldsymbol{A^T} * (\boldsymbol{y} - \boldsymbol{z}) = \boldsymbol{A^T} * (\boldsymbol{y} - \boldsymbol{A} * \boldsymbol{w}) = \boldsymbol{0} \tag{3.3}$$

If we solve this matrix equation for $\boldsymbol{w}$, then the vector 'closest' to $\boldsymbol{y}$ in subspace P is $\boldsymbol{z} = \boldsymbol{A} * \boldsymbol{w}$.

- We call $\boldsymbol{w}$ the **least squares approximation** to the linear system $\boldsymbol{Au} = \boldsymbol{y}$ in Equation (3.2). (Warning: it is not generally true that $\boldsymbol{Aw} = \boldsymbol{y}$.)

Equation (3.3) is expanded as

$$\boldsymbol{A^T y} - \boldsymbol{A^T A w} = \boldsymbol{0}$$

Rearranging the terms we have

$$\boldsymbol{A^T A w} = \boldsymbol{A^T y} \tag{3.4}$$

which we solve for $\boldsymbol{w}$. In fact, matrix $\boldsymbol{A^T A}$ is square and **if it is nonsingular** (invertible) then we have

$$\boldsymbol{w} = (\boldsymbol{A^T A})^{-1}\boldsymbol{A^T y} \tag{3.5}$$

which we call the least squares approximation of the linear system $\boldsymbol{Au} = \boldsymbol{y}$.

The matrix $(\boldsymbol{A}^T\boldsymbol{A})^{-1}\boldsymbol{A}^T$ plays such an important role in a variety of situations we give it a special name.

Definition: If $\boldsymbol{A}$ is an m × n matrix with linearly independent columns then $(\boldsymbol{A}^T\boldsymbol{A})^{-1}\boldsymbol{A}^T$ is called the **pseudo inverse** (or generalized inverse) of $\boldsymbol{A}$. It is denoted $\boldsymbol{A}^+$.

Thus the least squares approximation of $\boldsymbol{Au} = \boldsymbol{y}$ is given by

$$\boldsymbol{w} = \boldsymbol{A}^+\boldsymbol{y}$$

It is important to realize that we are not 'solving' the inconsistent system $\boldsymbol{Au} = \boldsymbol{y}$. Rather we replace the system $\boldsymbol{Au} = \boldsymbol{y}$ with the system $\boldsymbol{A}^T\boldsymbol{Aw} = \boldsymbol{A}^T\boldsymbol{y}$ which is easy to solve if $\boldsymbol{A}^T\boldsymbol{A}$ is nonsingular. To perform computations we need not repeat the arguments given above as to why this replacement generates an approximate solution. Instead we just use the result of the theoretical development. The linear system $\boldsymbol{A}^T\boldsymbol{Aw} = \boldsymbol{A}^T\boldsymbol{y}$ is called the **normal system** associated with the linear system $\boldsymbol{Au} = \boldsymbol{y}$.

In MATLAB the vector $\boldsymbol{w}$ in Equation (3.5) is obtained by typing the command

$$\mathbf{wlin} = \mathbf{inv}(\mathbf{A}' * \mathbf{A}) * \mathbf{A}' * \mathbf{y}$$

We will store the solution in **wlin** rather than just **w** so that it can be referred to later in this unit. Now in MATLAB compute the least squares approximation for the linear system in Equation (3.1), which corresponds to determining a linear model for the women's 100-meter dash. Show the MATLAB commands and their output in the space below.

In the space below write the least squares approximation to the women's 100-meter dash to construct a single linear equation of the form $y = ax + b$ that forms the linear geometric model to the data in Table 1.

$$y = \boxed{}\, x + \boxed{}$$

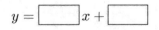

Next check your work by re-entering command **lsqline**, choosing input option 2, entering the names of the vectors containing the data, and inspecting the equation below the graph. You can see that the line does not go through all the ordered pairs of data from Table 1, but in some sense it comes close to all the data points. In fact, this line is the line that is closest to the 16 data points in the sense that the sum of the squares of the distances from the points to the line is minimized. (Thus we use the name **least squares approximation.**) Click on the Deviations On button and then on the Sum of Squares of Deviations. The vertical deviations will be displayed and the sum of the squares of the lengths of these line segments will be computed and displayed. The value is a (crude) measure of error incurred in using the least squares line as a model for this data.

Click on Evaluate the Model button and follow the on-screen directions. Set x = 2000 to predict the winning time in the women's 100-meter run for 2000. Record the value below.

Winning time predicted for 2000 by the Linear Model: ____________

In the space below explain how this winning time for 2000 was computed. Be precise.

How accurate is this prediction?

> Marion Jones of the United States won the event at the 2000 Olympics at Sydney in a time of 10.75. Thus the prediction of 10.68 (rounded to two decimal places) is off by only seven one-hundredths of a second.

What is the prediction for the 2012 Olympics? Record the result below.

Winning time predicted for 2012 by the Linear Model: ____________

This result is based on the data through 1996. Since we know the exact data for the 2000 Olympics, adjoin that information to the $\boldsymbol{x}$ and $\boldsymbol{y}$ vectors and run **lsqline** again. Exercise 1 asks for the result of this computation. Quit **lsqline**.

Exercises 2.

1. The data for this exercise is from the women's 100-meter dash. It can be loaded using command **w100dash**. Use **help w100dash** for details.

a) What is the equation of the least squares line for the using all the data from 1928 through 2000? Record the equation in the space below.

b) Use the equation in part a) to predict the winning time in the women's 100-meter dash in 2012. Record your result on the line below.

____

c) Use the equation in part a) to predict the winning time in the women's 100-meter dash in 2000. Record your result on the line below.

____

Does it match the exact winning time for 2000? Explain any difference in a short paragraph in the space below.

2. Let **A** = [**x ones(size(x))**] where **x** $= \begin{bmatrix} 1928 & 1932 & 1936 & 1948:4:2000 \end{bmatrix}'$.

a) Use the command **rref** to determine which columns of **A** are linearly independent. Write a short statement in the space below summarizing your findings.

b) Form the matrix $\boldsymbol{A' * A}$. In the space below explain how to show that it is nonsingular. Then verify your procedure.

Least squares models apply to many other Olympic events. Exercises 3 to 6 refer to a distance run, a high jump, a swimming event, and a long jump.

3. Here are the winning times (in seconds) for the men's 1500-meter run in the Olympics. (The data for this exercise can be loaded using command **m1500run**. Use **help m1500run** for details.)

year	time	year	time
1896	273.2	1956	221.2
1900	246	1960	215.6
1904	245.4	1964	218.1
1908	243.4	1968	214.9
1912	236.8	1972	216.3
1920	241.8	1976	219.17
1924	233.6	1980	218.4
1928	233.2	1984	212.53
1932	231.2	1988	215.96
1936	227.8	1992	220.12
1948	229.8	1996	215.78
1952	225.2	2000	212.07

a) What is the equation of the least squares line for the event for all years through 1996? Record your result below. (If you load the data with command **m1500run** delete the last value in both $\boldsymbol{x}$ and $\boldsymbol{y}$ for this part.)

b) Use the equation in part a) to predict the winning time in 2000 based on all Olympics through 1996. Record your result on the line below.

____

c) What is the equation of the least squares line for the event for all years through 2000? Record your result below.

d) Use the equation in part c) to predict the winning time in 2012 based on all Olympics through 2000. Record your result on the line below.

____

4. Here are the winning heights (in inches) for the men's high jump in the Olympics. (The data for this exercise can be loaded using command **highjump**. Use **help highjump** for details.)

year	height	year	height
1896	71.25	1956	83.5
1900	74.75	1960	85
1904	71	1964	85.75
1908	75	1968	88.25
1912	76	1972	87.75
1920	76.25	1976	88.5
1924	78	1980	92.75
1928	76.5	1984	92.5
1932	77.625	1988	93.5
1936	80	1992	92
1948	78	1996	94
1952	80.2	2000	92.5

a) What is the equation of the least squares line for the event for all years through 1996? Record your result below. (If you load the data with command **highjump** delete the last value in both $\boldsymbol{x}$ and $\boldsymbol{y}$ for this part.)

b) Use the equation in part a) to predict the winning height in 2000 based on all Olympics through 1996. Record your result on the line below.

____

c) What is the equation of the least squares line for the event for all years through 2000? Record your result below.

d) Use the equation in part c) to predict the winning height in 2012 based on all Olympics through 2000. Record your result on the line below.

____

5. Here are the winning times (in seconds) for the women's 100-meter freestyle swimming competition in the Olympics. (The data for this exercise can be loaded using command **w100free**. Use **help w100free** for details.)

year	time	year	time
1912	82.2	1964	59.5
1920	73.6	1968	60
1924	72.4	1972	58.59
1928	71	1976	55.65
1932	66.8	1980	54.79
1936	65.9	1984	55.92
1948	66.3	1988	54.93
1952	68.8	1992	54.65
1956	62	1996	54.50
1960	61.2	2000	53.83

a) What is the equation of the least squares line for the event for all years through 1996? Record your result below. (If you load the data with command **w100free** delete the last value in both $\boldsymbol{x}$ and $\boldsymbol{y}$ for this part.)

b) Use the equation in part a) to predict the winning time in 2000 based on all Olympics through 1996. Record your result on the line below.

____

c) What is the equation of the least squares line for the event for all years through 2000? Record your result below.

d) Use the equation in part c) to predict the winning time in 2012 based on all Olympics through 2000. Record your result on the line below.

____

6. The men's long jump raises an interesting issue because it contains one performance (by Bob Beamon in 1968) that statisticians call an 'outlier'. Here are the winning distances (in inches) for the men's long jump in the Olympics. (The data for this exercise can be loaded using command **longjump**. Use **help longjump** for details.)

year	distance	year	distance
1896	250	1956	308.25
1900	282.75	1960	319.75
1904	289	1964	317.75
1908	294.5	1968	350.5
1912	299.25	1972	324.5
1920	281.5	1976	328.5
1924	293	1980	336.25
1928	304.5	1984	336.25
1932	300.75	1988	343.25
1936	317.5	1992	341.5
1948	308	1996	334.75
1952	298	2000	336.75

a) What is the equation of the least squares line for the event for all years through 2000? Record your result below.

b) Use the equation in part a) to predict the winning distance in 2012 based on all Olympics through 2000. Record your result on the line below.

____

c) What is the equation of the least squares line for the event for all years through 2000 except 1968? Record your result below.

d) Use the equation in part c) to predict the winning distance in 2012 based on all Olympics through 2000 except 1968. Record your result on the line below.

____

7. The table below gives shoe sizes for women in the United States and in Europe.

x(U.S.)	$4\frac{1}{2}$	6	$6\frac{1}{2}$	7	$7\frac{1}{2}$	8	$8\frac{1}{2}$	9	$9\frac{1}{2}$	10
y(Europe)	35	37	$37\frac{1}{2}$	38	$38\frac{1}{2}$	39	40	$40\frac{1}{2}$	41	42

a) What is the equation of the least squares line. Record your result below.

b) Use the equation in part a) to determine the European equivalent of a size 12 shoe. Record your result on the line below.

____

8. Delta Airlines published a table showing how the temperature (in °F) outside an airplane changes as the altitude (in 1000 feet) changes.

x(Altitude in 1000's)	1	5	10	15	20	30	36.087
y(Temperature)	56	41	23	5	−15	−47	−69

a) What is the equation of the least squares line? Record your result below.

b) Use the equation in part a) to determine the temperature at 40,000 feet. Record your result on the line below.

____

9. The following table lists the number of doctorates awarded annually in mathematics to U.S. citizens and the percentage of doctorate recipients who are women. (The data for this exercise can be loaded using command **doctor**. Use **help doctor** for details.)

Year	U.S.	Women	Year	U.S.	Women
1973	774	10	1988	363	21
1974	677	9	1989	411	24
1975	741	11	1990	401	22
1976	722	12	1991	461	24
1977	689	13	1992	430	24
1978	634	14	1993	526	28
1979	596	16	1994	469	26
1980	578	15	1995	567	25
1981	567	18	1996	493	24
1982	519	17	1997	516	29
1983	455	20	1998	586	28
1984	433	20	1999	554	34
1985	396	20	2000	537	29
1986	386	21	2001	494	31
1987	362	20	2002	418	30

a) What is the equation of the least squares line for the number of U.S. citizens, with $x =$ years? Record your result below.

b) Use the equation in part a) to predict the number of U.S. citizens who will obtain a doctorate in mathematics in the year 2010. Record your result on the line below.

____

c) What is the equation of the least squares line for the percentage of doctorates who are women, with $x =$ years? Record your result below.

d) Use the equation in part c) to predict the percentage of women doctorates in the year 2010. Record your result on the line below.

____

<< NOTES; COMMENTS; IDEAS >>

<< NOTES; COMMENTS; IDEAS >>

Appendices

- Instructional Extensions to MATLAB
- Index of MATLAB Commands
- Index of Terms

<< NOTES; COMMENTS; IDEAS >>

Instructional Extensions to MATLAB

An important feature of this book is the set of instructional M-files which have been designed as aids for understanding fundamental linear algebra concepts. This section provides a list of the names of these commands together with a reference to the Lab in which they first appear. We also provide a description of each of these extensions to MATLAB . The descriptions are also available on-line through the **help** command.

Instructional Commands or Files [1]

alldesc		m1500run	LSQ	rowech	Lab 4
doctor	LSQ	mapcirc	Lab 13	rowop	Lab 2
evecsrch	Lab 13	matdat1	Lab 3	rrefquik	Lab 4
gschmidt	Lab 10	matdat2	Lab 3	rrefstep	Lab 4
highjump	LSQ	matops	Lab 3	rrefview	Lab 4
homsoln	Lab 6	matrixmaps	Lab 11	symrowop	Lab 2
igraph	Gr. Th.	matvec	Lab 13	symrref	Lab 4
invert	Lab 4	modn	Sec. Codes	uball	Lab 9
lincombo	Lab 7	planelt	Lab 11	vaultlsq	LSQ
lisub	Lab 6	project	Lab 10	vizrowop	Lab 2
longjump	LSQ	projxy	Lab 10	w100dash	LSQ
lsqgame	LSQ	rational	Lab 3	w100free	LSQ
lsqline	LSQ	reduce	Lab 2		

File **alldesc.txt** is an ascii text file that contains the collection of help files for each of the instructional M-files. It can be printed to have an easy reference document.

[1] Abbreviations: Gr. Th. - Introduction to Graph Theory; Sec. Codes - Secret Codes; LSQ - Least Squares Models

<< NOTES; COMMENTS; IDEAS >>

Index of MATLAB Commands [2]

The MATLAB commands used in this book together with the unit in which they are first introduced are displayed in the following table.

[2] We abbreviate Introduction to MATLAB and Some of its Features as Intro.

<< NOTES; COMMENTS; IDEAS >>

Index of Terms [3]

[3] MATLAB commands appear in bold face.

<< NOTES; COMMENTS; IDEAS >>